Ventilation

Other publications available from
Center for Occupational Hazards

Artist Beware: The Hazards and Precautions in Working with Art and Craft Materials
by Michael McCann, Ph.D., C.I.H.

Overexposure: Health Hazards in Photography
by Susan Shaw, M.P.H.

Health Hazards Manual for Artists
by Michael McCann, Ph.D., C.I.H.

Health and Safety in Printmaking
by Cherie Moses, James Purdham,
Dwight Bowhay, and Roland Hosein

Art Hazards Newsletter

Ventilation

Nancy Clark, M.A.
Thomas Cutter, P.E.
Jean-Ann McGrane, M.S.

NLB

Nick Lyons Books

Printed in the United States of America

10 9 8 7 6 5 4 3 2 1

Library of Congress Cataloging-in-Publication Data

Clark, Nancy.
Ventilation, a practical guide.

Reprint. Originally published: New York: Center for Occupational Hazards, 1984.
Bibliography: p.
1. Ventilation. 2. Artists—Health and Hygiene. 3. Artisans—Health and Hygiene. 4. Artists—Diseases. 5. Artisans—Diseases. I. Cutter, Thomas. II. McGrane, Jean. III. Title.
TH7678.C57 1987 697.9'2 86-27293
ISBN 0-94130-46-0

Figures were drawn by Sedonia Champlain unless otherwise credited.

Figure 13, page 51: This material is reproduced with permission from American National Standard Fundamentals Governing the Design and Operation of Local Exhaust Systems, ANSI Z9.2-1979, copyright © 1980 by the American National Standards Institute. Copies of this standard may be purchased from the American National Standards Institute at 1430 Broadway, New York, NY 10018.

Table 2, page 6: Reproduced by permission of the publisher from *Artist Beware*, copyright © 1979 Watson-Guptill.

Table 16, page 103: Reproduced from *1972 Fundamentals Volume, ASHRAE Handbook* (page 446), by permission of the American Society of Heating, Refrigerating and Air-Conditioning Engineers.

Table 17, page 104: Reproduced by permission of AVCO Industries.

Designed by Bruce McPherson

CONTENTS

Foreword

Judging from the number of inquiries received by the Art Hazards Information Center of the Center for Occupational Hazards about the subject, ventilation is one of the top concerns among artists today. Studio artists, craftspeople, public schools, art schools, community art centers, theaters, museums, small commercial shops—all want to make sure they are doing the right thing.

Their concern is justified. School inspections commonly reveal that existing ventilation systems often are inadequate or not working. Artists and museum conservationists who often need ventilation systems are found to have none at all, or the ones they have are the wrong types. Heating, ventilating and air conditioning engineers who have little experience and no training in industrial ventilation systems often design art-related systems, despite the need for industrial ventilation techniques, not commercial or domestic ventilation, in art situations.

Circumstances like these make the need for an art-related ventilation handbook obvious. Fortunately, panelists for the Design Art Program of the National Endowment for the Arts have agreed, and the Program's generous assistance has made the handbook possible. It is intended for artists, craftspeople, schools, universities, theaters, museums, small printing and woodworking shops—virtually anywhere art materials receive regular use. It should help those of you in these situations to evaluate your ventilation needs and help guide you through typical design steps without the aid of a professional engineer, and guide you also in deciding when situations are sufficiently complex to require professionals' aid.

How to use this book

Ventilation contains step-by-step procedures for designing and building your ventilation system. Part One, "Before you begin," will help you decide which processes in your workplace might require ventilation. Read this section before trying to decide which type of ventilation system you need. It discusses how to find out about the hazards of the chemicals you use and the precautions you might be able to take to decrease the cost of your ventilation system.

Part Two, "Introduction to ventilation," discusses basic ventilation concepts, types of ventilation systems, and how you can determine what type of ventilation system you need. It also describes in detail the components of ventilation systems, such as hoods, ducts, fans, and air cleaners. And it tells you how to maintain your ventilation system and where to find experts you need for design and installation.

Part Three, "Five specific ventilation-system designs," gives specifications for five standard types of ventilation systems: canopy hood, slot hood, spray booth, woodworking dust-collection system, and movable exhaust system. These specifications provide general guidelines for designing specific systems. In many instances, the calculations illustrated in Appendix B will enable you to modify these general designs to suit your individual situations. You may need to have your specific plans verified by a professional engineer or a heating, ventilating and air conditioning engineer who is experienced in designing industrial ventilation systems.

This handbook also contains several useful appendices about applicable codes and regulations, sample design calculations with essential charts, a list of suppliers, a bibliography, and a glossary.

If you follow the steps and heed the caveats, you should be able to demystify ventilation and come up with a system suited especially to your needs.

Michael McCann, Ph.D., C.I.H.
Executive Director
Center for Occupational Hazards

I

Before you begin

1 Know your materials

Once most people decide they need a ventilation system, they want it immediately. The impulse to hurry is natural. But as with most major purchases, it is wiser to take your time, plan, and familiarize yourself with some basic principles before spending your money.

Here are some considerations to help you evaluate your studio or work setting, together with tips about gathering the information you need to choose the right system. The time you spend planning now will save you even more time and money, and possibly your health, in the future.

The first consideration is the materials you work with. What are they? Do they contain hazardous ingredients? How serious a health problem do they pose for you, your co-workers, or your family? What actually is in the materials? It is unfortunate but true that most people do not know what is in the materials they use, nor the potential health effects of the ingredients. All too often the consumer relies upon an assumption that products sold nowadays are "safe." Distributors, local art suppliers, and hardware store owners usually reassure purchasers when they express concern. But such reassurances really are not enough.

Virtually everyone who undertakes product research independently soon finds there is much to learn about the health hazards of chemicals. The investigator finds, too, that few people who sell products know what the products' hazards are. The best advice is to find out for yourself.

Your principal concern is to determine how toxic or physically harmful a chemical or its ingredients are. Some important points about evaluating chemical hazards are reviewed in this section. But for a more detailed discussion, read the reference works listed in the bibliography at the end of the book.

Chemical toxicity

The word "toxicity" is a way of describing a material's ability to damage the body. The higher its toxicity, the smaller the amount of the material needed to cause damage. Highly toxic materials require special handling procedures and ventilation controls (to be discussed later). Remember, *low concentrations of highly toxic materials can still do serious, permanent harm*. Do not assume, as so many persons have, that using only a little bit of a highly toxic chemical makes it OK or safe. A little bit can pack a lethal punch.

Chemicals' health effects are either *acute* or *chronic*.

An acute effect usually results from a single exposure which occurs in one of three ways:

Ingestion—for example, mistakenly drinking turpentine which had been stored carelessly in an empty soda bottle

Inhalation—breathing air with high carbon monoxide levels from an improperly vented gas kiln

Skin contact—receiving acid burns from spills during an etching process

The general rule about acute health effects is that the reaction is usually immediate and the cause is usually apparent.

A chronic effect, on the other hand, is a long-term health problem, usually resulting from low-level exposure to chemicals over an extended period. These can include chronic lung problems, blood vessel and heart damage, liver and kidney impairment, and other serious problems. For many chronic diseases such as silicosis (a debilitating lung disease caused by exposure to clay and other silica containing dust) or liver damage (as from continuous solvent exposure), years or even decades may pass between the time of initial chemical exposure and the time when symptoms develop. This time gap is called the latency period. Because of the latency period and the fact that such chronic diseases are often very serious, there is frequently little that can be done for the patient by the time symptoms appear.

Cancer can be a chronic health effect of exposure to certain chemicals, and it deserves special attention. Exposure to particular chemicals can cause specific types of cancer: lung cancer from inhaling arsenic oxide, or lead or zinc chromate, uranium oxide, nickel compounds, cigarette smoke, or asbestos; leukemia from exposure to benzene (benzol); bladder cancer from exposure to benzidine-based dyes; or skin cancer from synthetic cutting oils. Like other chronic diseases, cancer may take from ten to forty years to develop. *At present there is no known safe exposure level to cancer-causing materials.* The safest course is to avoid all exposure.

A material's *form* also determines how much will get into the body and be absorbed. The major forms chemicals take are dusts, fumes, mists, vapors, and gases. A description of these various forms or categories appears in Table 1.

The particulates—dusts, fumes and mists such as paint spray mists—differ from one another in their size and ability to remain in the air. The smaller and lighter the particle, the more likely it is that it will remain suspended in the air for hours. These smaller particles easily penetrate into the deepest parts of the lungs, and they can remain in the lungs or be absorbed into the body. Gases and vapors also easily penetrate into the deep parts of the lungs.

Researching chemical hazards

Finding out what is in products usually takes some effort. Container labels on most products used by artists customarily provide little useful information about product ingredients. Warnings like, "Caution, This Product May Be Harmful if Inhaled," or, "Use in Well Ventilated Areas," do not tell you how to protect yourself or how to provide adequate ventilation. The name on a container is often a *trade name*—the name used by the supplier—and not the *chemical* or *generic* name. For example, Varsol is a trade name for mineral spirits. To determine chemical toxicity, you need to know the generic or chemical names of *all* the ingredients in a product. For most products, this means you must start by obtaining a *Material Safety Data Sheet.*

Figure 1 consists of a commonly used Material Safety Data Sheet (MSDS) and an explanation of the sheet. When you design a ventilation system to handle specific compounds or chemicals, pay special attention to Section VIII (or its equivalent on other forms), which provides guidelines about the types of ventilation systems to consider.

Table 1

Physical Forms of Materials

DUSTS	Particles with a wide range of sizes and shapes; generated by handling, crushing, grinding (such as clay or wood dust)
MISTS	Droplets which become airborne through bubbling, boiling, spraying, splashing, or other agitation; also formed from liquids at room temperature (such as acid mist or aerosol spray can mist).
FUMES	Small, solid particles formed above molten metal, generally after volatilization, or by a chemical reaction like oxidation (also welding or soldering fumes, for example).
VAPORS	Gaseous forms of substances which are normally in liquid or solid state; changed to these states by increases in pressure or decreases in temperature (like solvent vapors).
GASES	Compressible formless fluids which occupy a space or enclosure.

Source: National Institute for Occupational Safety and Health. *Industrial Environment, Its Evaluation and Control.* (Washington D.C.: U.S. Government Printing Office, 1973).

Figure 1 Guide to the Use of Material Safety Data Sheets

Section I. Identifies the product and its manufacturer.

Section II—Hazardous Ingredients. Lists almost all product ingredients for which OSHA has a standard, along with their Threshold Limit Values (TLVs).

Section III—Physical Data. Lists certain chemical data—but no health hazards. Can be useful in deciding how dangerous a product is.

Section IV—Fire and Explosion Data. Gives the flash point of the material—the temperature at which the material will ignite in the presence of a spark or flame. If the flash point is near or below 100°F., the material is particularly dangerous—hot weather, static electricity, or a cigarette can set off a fire or explosion.

Section V—Health Hazard Data. Often gives poor quality information. Short-term hazards may be listed, but not long-term ones such as cancer.

Section VI—Reactivity Data. Gives information on conditions that could cause the product to react dangerously or to decompose and release dangerous materials.

Section VII—Spill or Leak Procedures. Gives instructions for handling spills.

Section VIII—Special Protection Information. Lists protective equipment and proper ventilation to be used with the material—probably the *minimum* measures.

Section IX—Special Precautions. Lists handling, storage, and other special safety measures.

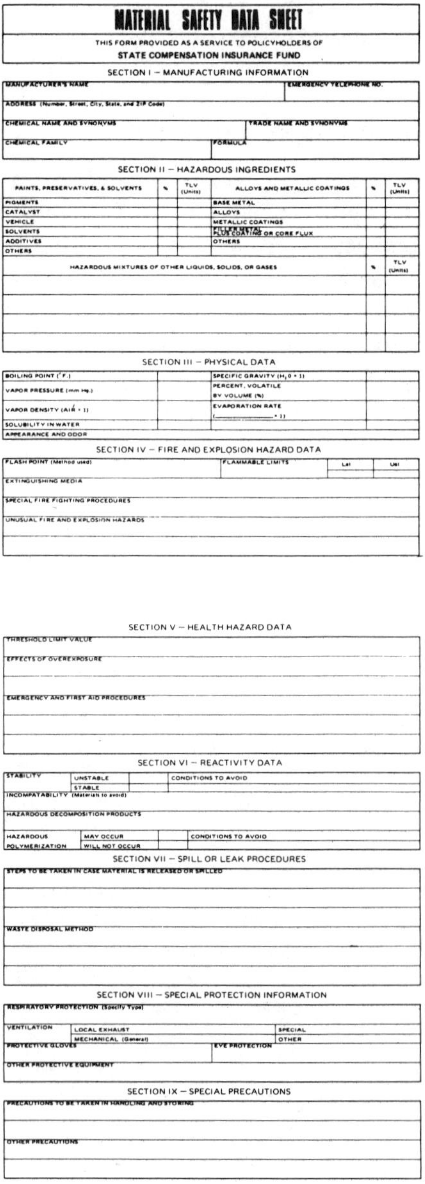

MATERIAL SAFETY DATA SHEET

THIS FORM PROVIDED AS A SERVICE TO POLICYHOLDERS OF
STATE COMPENSATION INSURANCE FUND

SECTION I – MANUFACTURING INFORMATION

MANUFACTURER'S NAME | EMERGENCY TELEPHONE NO.
ADDRESS (Number, Street, City, State, and ZIP Code)
CHEMICAL NAME AND SYNONYMS | TRADE NAME AND SYNONYMS
CHEMICAL FAMILY | FORMULA

SECTION II – HAZARDOUS INGREDIENTS

PAINTS, PRESERVATIVES, & SOLVENTS	%	TLV (Units)	ALLOYS AND METALLIC COATINGS	%	TLV (Units)
PIGMENTS			BASE METAL		
CATALYST			ALLOYS		
VEHICLE			METALLIC COATINGS		
SOLVENTS			FILLER METAL PLUS COATING OR CORE FLUX		
ADDITIVES			OTHERS		
OTHERS					

HAZARDOUS MIXTURES OF OTHER LIQUIDS, SOLIDS, OR GASES	%	TLV (Units)

SECTION III – PHYSICAL DATA

BOILING POINT (°F.)		SPECIFIC GRAVITY (H_2O = 1)	
VAPOR PRESSURE (mm Hg.)		PERCENT, VOLATILE BY VOLUME (%)	
VAPOR DENSITY (AIR = 1)		EVAPORATION RATE (______ = 1)	
SOLUBILITY IN WATER			
APPEARANCE AND ODOR			

SECTION IV – FIRE AND EXPLOSION HAZARD DATA

FLASH POINT (Method used) | FLAMMABLE LIMITS | Lel | Uel
EXTINGUISHING MEDIA
SPECIAL FIRE FIGHTING PROCEDURES
UNUSUAL FIRE AND EXPLOSION HAZARDS

SECTION V – HEALTH HAZARD DATA

THRESHOLD LIMIT VALUE
EFFECTS OF OVEREXPOSURE
EMERGENCY AND FIRST AID PROCEDURES

SECTION VI – REACTIVITY DATA

STABILITY | UNSTABLE | STABLE | CONDITIONS TO AVOID
INCOMPATABILITY (Materials to avoid)
HAZARDOUS DECOMPOSITION PRODUCTS
HAZARDOUS POLYMERIZATION | MAY OCCUR | WILL NOT OCCUR | CONDITIONS TO AVOID

SECTION VII – SPILL OR LEAK PROCEDURES

STEPS TO BE TAKEN IN CASE MATERIAL IS RELEASED OR SPILLED
WASTE DISPOSAL METHOD

SECTION VIII – SPECIAL PROTECTION INFORMATION

RESPIRATORY PROTECTION (Specify Type)
VENTILATION | LOCAL EXHAUST | MECHANICAL (General) | SPECIAL | OTHER
PROTECTIVE GLOVES | EYE PROTECTION
OTHER PROTECTIVE EQUIPMENT

SECTION IX – SPECIAL PRECAUTIONS

PRECAUTIONS TO BE TAKEN IN HANDLING AND STORING
OTHER PRECAUTIONS

Source: MassCOSH Fact Sheet.

Even though Material Safety Data Sheets are useful, they have some limitations. For example, often only those materials with Occupational Safety and Health Administration (OSHA) standards or low flash points are listed under Section II which is concerned with hazardous ingredients. Also, for example, such long-term effects as cancer or reproductive damage are usually not even mentioned in Section V. Despite such limitations, MSDS's are worthwhile guidelines. Use them as a starting point for further research.

Knowing the potential hazards of individual chemicals is essential. But it is not enough. Mixing or heating some chemicals can cause chemical reactions that produce toxic by-products. For example, etching zinc or copper with nitric acid produces toxic nitrogen dioxide gas. Or welding near a chlorinated solvent such as trichloroethylene can produce deadly phosgene gas. So, if you mix or heat chemicals, be sure to learn their potential breakdown or decomposition products when you conduct your hazard evaluations.

Fire hazards

Organic solvents and other liquids used by artists and craftspeople can create serious fire hazards. Storage and handling of such flammable liquids is strictly regulated by local and federal laws. (See Appendix A for a discussion of codes and regulations.) Even if you use only small amounts of flammables, you should be familiar with how to handle them safely and with proper work practices. Fires and explosions of flammable liquids can cause serious injuries.

Liquids become a potential fire hazard when they vaporize. Such vapors can be ignited when they concentrate in sufficient volume above a liquid's surface. A liquid's seriousness as a fire hazard depends upon its flash point. The flash point is the lowest temperature at which vapors, collected in sufficient quantity above a liquid's surface, form an ignitable mixture in the air. A liquid with a low flash point can generate an explosive atmosphere at or below room temperature. One with a high flash point would not pose a serious hazard unless heated.

Once an ignitable mixture is formed in the air, any number of ignition sources can start a fire or touch off an explosion. Common ignition sources which you should guard against in your studio or workshop are cigarettes, matches, welding torches, sparking tools, and sources of static electricity.

Depending upon their flash points, liquids are grouped into three categories: nonflammable, flammable, or combustible. Each category requires specific handling, storage, and ventilation precautions.

Flammability also has been categorized. And to complicate matters somewhat, two different categorization systems exist in the United States. One, adopted by the National Fire Protection Association (NFPA), covers most *industrial* materials. The second, defined by the Federal Hazardous Substances Act (FHSA), covers *consumer* products bought in art supply and hardware stores, including paint strippers, aerosol sprays, and thinners. Therefore, you have to be familiar with both systems.

In the industrial system, the NFPA divides flammable liquids into three categories: Class I Flammable if the flash point is below 100°F, and Class II or III Combustible if the flash point is above 100°F. NFPA further divides Class I Flammables into IA, IB, and IC. Class IA and IB liquids have flash points below normal room temperatures. These liquids (often solvents) are extremely hazardous because they can start flash fires. Class IC flammable liquids are potential fire hazards on hot days or when they are kept near a source of heat.

The combustible liquids with flash points above 100°F are divided into Classes II, IIIA, and IIIB. Although any combustible liquid is a fire threat if it is heated, Class IIIB liquids generally cause little concern as a fire hazard because of their high temperature flash points.

The FHSA definitions for flammability and combustibility differ from the NFPA's. The most dangerous flammable liquids are those labeled "extremely flammable." Common examples are acetone and rubber cement thinner. These require special precautions. FHSA definitions appear in Table 2.

The differences between the NFPA and FHSA definitions involve liquids with flash points between 80°F and 100°F. The NFPA defines these liquids as *flammable,* Class IC, whereas the FHSA defines them as *combustible*. Another difference is that NFPA categorizes as IC those liquids that can reach their flash points on a hot summer day or in the presence of a heat source that affects room temperature. The solvents falling into this category are turpentine, xylol, styrene (polyester resin), and some lacquer thinner components.

Which guidelines should be used in studios and workshops? Because the National Fire Protection Association definitions for industrial use provide a more protective setting, they should be used in occupational and professional environments. Table 3 lists some common solvents and their NFPA categories.

Explosive limits of vapors and gases

Flammable or combustible vapors mixed with air can explode if they fall within a certain range of concentrations. The minimum concentration of vapors

Table 2

Federal Hazardous Substances Act: Labeling Requirements

Hazard category	*Flash point*
Extremely flammable	below 20° F (−7° C)
Flammable	20° to 80° F (−7° C to 27° C)
Combustible	80° to 150° F (27° C to 66° C)

Table 3

NFPA Flammability and Combustibility Definitions

FLAMMABLE LIQUIDS		
Class IA Flash point: below 73° F (23° C) Boiling point: below 100° F (38° C)	**Ethyl ether** **"Flammable" aerosol sprays**	
Class IB Flash point: below 73° F (23° C) Boiling point: over 100° F (38° C)	**Acetone** **Benzine** (VM&P naphtha) **Benzol** (benzene) **Butyl acetate** **Cyclohexane** **Dioxane** **Ethyl acetate** **Ethyl alcohol**	**Ethylene dichloride** **Gasoline** **Hexane** **Isopropyl alcohol** **Methyl acetate** **Methyl alcohol** **Methyl ethyl ketone** **Toluol** (toluene)
Class IC Flash point: 73°–100° F (23°–38° C) Boiling point: over 100° F (38° C)	**Amyl acetate** **Amyl alcohol** **Butyl alcohol** **Methyl butyl ketone** **Methyl isobutyl ketone** **n-Propyl alcohol** **Styrene** **Trichloroethylene** **Turpentine** **Xylol** (xylene)	
COMBUSTIBLE LIQUIDS		
Class II Flash point: 100°–140° F (38°–60° C)	**Cellosolve acetate** **Cyclohexanone** **Dimethylformamide** **Ethyl silicate** **Isoamyl alcohol** **Kerosene** **Methyl cellosolve** **Mineral spirits** (odorless paint thinner)	
Class IIIA Flash point: 140°–200° F (60°–93° C)	**Butyl cellosolve**	
Class IIIB Flash point: over 200° F (93° C)	**Cellosolve** **Diethylene glycol** **Ethylene glycol** **Hexylene glycol**	

Source: Michael McCann. *Artist Beware* (New York: Watson-Guptill, 1979). Reproduced by permission of the publisher.

in air that will ignite is called the lower explosive limit (LEL) and also the lower flammability limit (LFL). Likewise, the maximum concentration over which no explosion will occur is called the upper explosive limit (UEL) or upper flammability limit (UFL). A mixture of flammable vapors and air will ignite at all concentrations between the LEL and the UEL. Of course, an ignition source is always needed to cause an explosion.

To guard against fire and explosion, the concentration of flammable and combustible liquids in air (especially those with flash points below 100°F) should be kept well below ten percent of the LEL.

As a general rule, any liquid which is being controlled as a health hazard automatically is being controlled as well as a fire hazard. The LEL of a solvent usually is much higher than the concentration that affects health adversely. For example, xylol has a health hazard rating (Threshold Limit Value or TLV) of 100 parts per million (ppm) in air. But its LEL is one percent or 10,000 ppm; ten percent of the LEL is 1,000 ppm. If the concentration of xylol is kept below its TLV of 100 ppm to control it as a health hazard, its potential as an explosion hazard also is well under control.

Explosion hazards of dusts

Liquids are not the only fire and explosion hazards. Heavy concentrations of pulverized, fine organic (and some metal) dusts belong in that category, too. If the dust concentration is great enough, an inadvertent spark from a faulty motor, or from two pieces of iron or steel being rubbed together, or from numerous other sources can touch off an explosion. The *quantity* of the dust does not have to be large. It is the dust's *concentration* that makes the difference. If the process which creates the dust is conducted in an enclosure of some sort, and if that enclosure permits dust to build up, an explosion hazard may result.

Dust explosions occur frequently in such places as grain elevators and bakeries. But they also can occur in a studio. For example, take the buildup of rosin in rosin boxes used for aquatint in intaglio. The box allows the rosin dust to concentrate. If static electricity or a spark can be generated, an explosion may occur. Therefore, all parts of the rosin box should be non-sparking.

Flammable materials require very special consideration in ventilation system design. A liquid's flash point, its concentration, and the work process temperature will determine if special precautions to avoid fire or explosions will be necessary.

Once you have evaluated the total hazard potential of your system, make a special list of any highly toxic or highly flammable materials on it. If possible, find substitutes for these materials. You should do this *before* designing the system, not *after*. Safer substitutes will decrease your health risk, and they very likely will reduce the cost of the system as well.

2 Evaluation of process

At this point, you should have developed a complete inventory of your chemicals and you should be familiar with the types of hazards each presents. The next step is to review carefully your actual work processes and the setting where the chemicals will be used. These facts and details will help you choose an appropriate ventilation system for your workspace.

Form of the material

The form of the material used or produced during the process (dust, mist, vapor, gas, fume) will affect your overall system design. Any process that involves extreme temperatures or corrosive materials (such as acids or alkalies) will require special consideration when you choose the materials to construct your system. You should consider non-flammable and non-corrosive components to insure safety and a long system life.

The form of the materials you work with also will determine the kinds of hood and fan your system will need. For example, moderate to large dust particles, such as wood dust from planing or sawing, will require one type of system; fine dust, such as wood dust from machine-finish sanding, will require a different type. Solvent vapors generated from a silk screen process can be captured effectively in a slotted hood, but hot air and kiln gas emissions will require an overhead or canopy hood. The various kinds of ventilation systems required for different processes are discussed in Part II.

Exposure: How much is too much?

Assessing the length of time you work at one stretch and the frequency of your work may help you choose among several alternative ventilation systems. For example, if you work on a project only once a year for one day, and if you do not use any highly toxic material, some kind of *dilution ventilation* may be your proper choice. But if your work involves a continuous operation with high levels of exposure to a material of low toxicity, or a continuous operation with low-level exposure to a highly toxic material, you will need a *local exhaust system*.

Recommended standards as guidelines

Determining high or low exposure levels for individual chemicals may be difficult, and it may not be necessary for your particular situation. Many professionals who design ventilation systems (ventilation engineers or industrial hygienists) use as their basic reference tools two sets of standards for measuring chemical exposure.

One is the time-weighted average Threshold Limit Values set by the American Conference of Governmental Industrial Hygienists (ACGIH). The other is the Permissible Exposure Limits set by the Occupational Safety and Health Administration (OSHA). Both sets are based upon average eight-hour exposure levels to particular chemicals. However, these eight-hour average standards cannot be used for chemicals that have high acute or immediate toxicity or which are highly irritating. For these chemicals, either Short Term Exposure Limits or Ceiling Threshold Limit Values are used. Short Term Exposure Limits are based on average concentrations measured during a fifteen-minute period; ceiling limits are concentrations which should never be exceeded, even momentarily.

The ACGIH revises its TLVs regularly, and they generally give more protection than the PELs set by OSHA. However, the ACGIH is an advisory group of professionals, and the organization's standards are therefore only recommendations. OSHA, on the other hand, is a federal agency with legal powers to set and to enforce standards. OSHA standards are the law.

OSHA standards for chemical exposure are listed in Section 1910.1000 of the OSHA General Industry Standards. A copy can be obtained from the OSHA office in your area.

Even though many professionals use OSHA standards when they design a ventilation system, these standards have many weaknesses which must be considered. For example, most OSHA standards are based upon 1968 TLVs. Most of these 1968 TLVs were keyed to acute, not chronic, effects of toxicity. This means that most of the standards do not reflect health effects such as cancer, reproductive damage, or hard-to-pinpoint generalized illnesses such as fatigue, headaches, or slowed nerve-conduction response time. Then too, for many chemicals there are no standards. And besides this, both TLVs and PELs were set for *individual* chemicals, so they provide no guidance about safe exposure levels to *mixtures* of several chemicals.

When you design your own ventilation system, probably the most important thing to remember is that TLVs and PELs were not established to protect particular persons. They were set for the "average" person. Certain groups such as the elderly, children, asthmatics, or other high-risk or hypersensitive groups of people were not taken into consideration. They therefore are not protected at the levels set by the standards.

Rather than worry about specific chemical levels, you probably should attempt simply to control exposure at the lowest level possible. You can attain this goal feasibly if your ventilation system is designed properly.

3 Work practices often overlooked

A ventilation system is only one specific part of a total effort to provide a healthy and safe workplace. Equally important are good work practices—including proper storage, material containment, use of personal protective equipment, and good housekeeping. Good work practices also can help lower a ventilation system's cost and improve its efficiency. The tips that follow are quite simple yet they often are overlooked in workplaces of all sizes and types.

Good housekeeping

Good housekeeping is essential in any workplace. You should clean floors or work surfaces daily, and clean up toxic spills immediately to decrease the potential for skin contact and for inhaling the contaminant in the workroom air. Toxic dusts, for example, should be removed by using either an appropriate vacuum cleaner or a wet mop. Dry sweeping puts dust into the air and increases inhalation potential. Getting into the habit of cleaning up your workspace will help you ultimately to establish a routine maintenance program—and that program is essential to keeping your ventilation system in good working order in the future.

Most household vacuum cleaners are dangerous when used with toxic dusts since they pass fine particles through their filters. These then become airborne and can be inhaled. Special—not household—vacuum cleaners, however, are very effective for cleaning up accumulated wastes in workspaces. In fact, vacuum cleaning is one of the two recommended clean-up methods for toxic dusts. The other is the wet-down method, where dusts are wetted and then swept or vacuumed. Good housekeeping not only keeps workspaces clean and easier to work in, but it keeps them safe by preventing the return of dusts to the air and by reducing the amount of work that a local exhaust ventilation has to do.

Three vacuum-cleaning methods are useful in a small shop or studio: a high-efficiency particulate air (HEPA) filter vacuum cleaner, a wet pick-up vacuum cleaner, and a wet vacuum cleaner.

For highly toxic materials, such as lead, silica, ceramic dusts, and asbestos, HEPA vacuums or wet pick-up vacuums are the only recommended types. Both can pick up very fine particles. The HEPA filter vacuum works by drawing contaminated air through a high-efficiency filter which traps particles of microscopic sizes. The wet pick-up vacuum works in a two-step process: first, you wet down the dusty area with plenty of water, then you use an industrial vacuum to pick up the slurry. Both vacuums return cleaned air to the work area. You must dispose of the collected contaminants properly.

The third kind, the wet vacuum cleaner, is an industrial vacuum cleaner that uses a water reservoir to collect heavy particles from the airstream. As the air passes over the water, the particles settle on its surface. The filter does not clog readily, and the motor does not have to pull air through a highly resistant filter. Thus, filters are not changed as often and the motor is protected against burnout. A wet vacuum is good only for medium-sized particles and never should be used for highly toxic materials.

Limiting use of toxic materials

Limiting the use of toxic materials to as small an area as possible will help decrease your housekeeping problems and will limit the size and the complexity of your ventilation problem. For example, if you measure and mix a powdered dye in one room, then heat the dye baths in another room, the two separate areas must be considered when you design a ventilation system. But if you can mix, measure, and heat the dyes all in one area, then you can much more easily and efficiently vent the area, control access to the dyes, and accomplish your housekeeping chores.

You can limit exposure either by organizing the work process, so that the use of toxics is confined to one specific area, or by enclosing certain operations. Specialists prefer enclosure if you work with highly toxic materials, especially *carcinogens*. In any case, enclosure is useful for many operations.

Suppose you need an enclosure to mix powdered dyes for batik work. When used continually, some of these dyes can cause serious health problems. For example, the cold-water, fiber-reactive type can cause severe respiratory allergies and asthma. When you mix small to moderate quantities of dyes, you can use a simple glove box to enclose the process, decrease airborne exposure, and limit your housekeeping problem. To make a glove box, select a cardboard box, cut two holes in the side for your hands, shellac the box on both sides, and install a piece of Plexiglas on top for viewing. Then you can mix dyes with very little exposure. After mixing, you can remove the Plexiglas top and wipe the box with a damp cloth to remove any spilled dye powder.

Figure 2 Enclosure for Mixing Powders

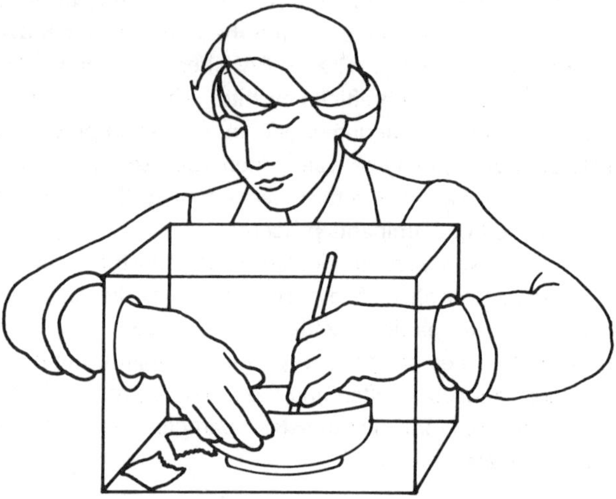

Safe storage and handling of materials

Safe storage and handling of materials used are essential for controlling accidental chemical exposures. Select storage containers made of materials which will not react with the chemical (hydrofluoric acid should be kept in special plastic and not in glass containers, for example) and which will be strong enough to withstand normal use. Never store near each other any chemicals that could react with each other, such as cyanide and acid, or methyl ethyl ketone peroxide and acetone. Also, when you store chemicals, avoid hard-to-reach areas where you will increase the likelihood of spills or broken containers. If you use large quantities of flammable and combustible materials, you are required to store them in special cabinets. Check storage regulations if you work with these materials.

You should cover solvents and powders. Covers will prevent unanticipated reactions from spills, eliminate fire hazards, and reduce evaporation.

Labeling of chemicals in your workplace

All chemicals in your workplace should be labeled with contents and hazards, if known.

Remember, although you may know what the chemical in the empty juice container is, someone else may not.

Good personal hygiene

Good personal hygiene can decrease chemical exposure significantly. Do not allow smoking and eating in the workplace. MAKE NO EXCEPTIONS! If you work with powders, you should wear special clothes and keep them in the workplace. These clothes should be washed separately from the clothes of others in the family or others with whom you work. If you send such clothes to a laundry, warn the staff that the clothes contain toxic materials and should be handled specially.

If chemicals do splash or spill on anyone, wash the exposed area immediately. Delay may result in serious long-term damage.

Personal protective equipment

Wearing personal protective equipment should not be necessary if your ventilation system is designed correctly and you follow good work practices. However, there are exceptions. A material's toxicity and the potential route of exposure are two considerations which will determine the need for personal protective equipment. You will need impermeable gloves and aprons for anything that may be absorbed through the skin. Wear goggles to guard against materials that can cause eye irritation, such as acids or alkalies, abrasives, or flying particles such as metal or wood from grinding or chipping operations. Finally, you may need a respirator to eliminate inhalation exposure from those processes that are not served by the new ventilation system. Choosing a respirator is not the simple task many persons think it is. All respirators are not alike, and not all respirators should be used for all purposes. *Artist Beware* by Michael McCann provides further information on the selection of a respirator.

Safe chemical disposal

Safe chemical disposal will further limit personal and community exposure. State and federal governments regulate disposal of certain extremely toxic chemicals such as PCBs (arochlors), asbestos, or pentachlorophenol. For other chemicals, such as lead, uranium, or benzene, local disposal regulations may apply.

In general, do not pour solvents down the sink. Most sewage treatment plants (and septic tank systems) are not designed to absorb organic solvents, which can get into the system and kill the bacteria needed to break down other waste products.

Most unregulated aqueous liquids can be poured down the sink singly with lots of water to dilute their concentration. Any unregulated material which is placed in the regular trash should be packaged so that spills will not occur and so that sanitation workers or other persons will not be exposed to it.

4 Looking at the work area

Unfortunately many artists and people who own or run small businesses face strict limits in their choices of workspace. However, if you have any choice at all, you may find the following basic points a useful guide.

Location of work

1. *Do not mix work and living spaces.* Do not select a work area where other family members eat or play. Too often young children have access to parents' work areas. Toddlers can easily ingest materials (toxic dusts such as lead, for example) when materials accumulate on floors or other surfaces.

2. *Understand the limits of your space and your finances.* Whether you own or rent your workspace may affect how much renovation you are willing to undertake and how much money you are willing to spend. If you foresee moving in the near future, add the portability of the ventilation system to the questions on your list, Building codes and insurance regulations also may affect your choice. Check these codes and also the appropriate sections of your lease to determine what requirements must be met in your building. Refer to Appendix A for a discussion of codes and regulations.

3. *Limit the distance the air must travel in the ventilation system.* If you enjoy some flexibility in deciding where to place your workspace, you should keep in mind several general principles of air flow.

Visualize a volume of air as a cardboard box. Then imagine moving that box. You can see that the size of the box as well as the distance it has to travel are critical factors. Air is no different. The larger the volume of air and the greater the distance it must be moved, the more energy is required to move it.

In a ventilation system, distance and energy translate into money. Galvanized ducts, for example, may cost $10 to $25 a foot. The longer the duct that will carry toxic materials from the worksite to the outside, the greater the cost.

Figure 3 How to review the safety of a process

1. Make a diagram of your workspace. Indicate location of work areas; windows, doors, outside walls; chemical storage areas, mixing, drying, and processing areas; locate sources of contaminated air.

2. For each material used or stored, provide this information.

Trade name__________

Chemical name of ingredients__________

Material Safety Data Sheet available[a]__________

Physical form[b]__________

Flammability hazard[c]__________

Toxicity hazard[d]__________

3. Rate each work practice "good" or "poor" and list changes needed.

	Rating	Changes
Housekeeping	____	________
Personal protective equipment (respirators, gloves, clothing, etc.)	____	________
Disposal or storage of materials	____	________
Isolation from living area	____	________

4. Review legal and contractual requirements of lease, insurance policies, OSHA, local environmental and building laws.

5. Develop plan to reduce exposure. Consider changing machinery or equipment needing ventilation, relocating the workspace, or installing a ventilation system.

a. Material Safety Data Sheets should be available from the manufacturer.

b. Note the physical form of material (dust, fume, gas, vapor, etc.)

c. Use the Federal Hazardous Substances Act definitions for flammable materials:

EF, extremely flammable, for a substance with a flash point below 20° F (−7° C);

F, flammable, for a substance with a flash point between 20° and 80° F (−7° and 27° C); or

C, combustible, for a substance with a flash point between 80° and 150° F (27° and 66° C).

d. Use TLV for toxicity rating:

Slightly toxic: TLV ≥ 500 ppm or ≥ 10 mg/M^3

Moderately toxic: TLV 100–500 ppm, or 1–10 mg/M^3

Highly toxic: TLV ≤ 100 ppm or ≤ 1 mg/M^3

Keeping these things in mind, plan your work area near a window or at least near an outside wall. (Certain exceptions will be discussed later.) The closer the worksite to the point where the air exits, the lower the cost will be for removing the contaminated air. A window is preferred because it is an already available opening in a wall through which air can be exhausted directly or indirectly.

Process review guide

At this point, you should have identified the potential hazards of the chemicals with which you work, and you should have reviewed your work process to prepare for selecting a ventilation system. The following process review guide will help you organize and review this information (Figure 3).

II

Introduction to ventilation

5 Basic principles

Ventilation systems are designed to move air in, around, and out of rooms. Some systems move air in order to provide a comfortably cooled or heated environment. Other systems, like the ones described in this book, move air in order to remove airborne contaminants from a work environment.

Whatever a ventilation system's purpose, controlling air movement is no simple matter. Air blowing through open windows and doors can cool a room if the wind blows from the right direction. But if the goal is more complex, such as removing solvent vapors, metal fumes, or other toxins from the air, then an open window or door cannot provide "adequate ventilation." A ventilation system is needed. And it must be designed to draw the contaminated air away from the breathing zones of people in the room and carry it safely outside.

To ventilate your shop or studio in order to keep the air healthy, you need to know what makes air move, how much air movement is needed, and what restricts air movement. Understanding basic air flow principles will help you select a ventilation design, install it (or have it installed) properly, and maintain it so that the system can do what it is supposed to do—keep the air fit to breathe.

Air moves naturally from one place to another because of differences in air pressure. Air molecules in a high (or positive) pressure area will flow automatically to a low (or negative) pressure area. Air movement always occurs along the path of least resistance. In a ventilation system, a fan causes air to flow by creating a difference in pressure.

Air resistance in ventilation systems hinders the flow of air. Resistance is caused by friction—air rubbing against itself or the surfaces of ventilation components—and by turbulence, which is haphazard movement caused by the air's changing speed and direction.

It is hard to visualize air flow because for the most part air is invisible. So imagine a morning rush hour in a busy city. People leave their homes and head for the subway. Some stroll along, some run to catch the train, some walk together in groups, and others walk alone. For the most part, people move without much resistance. But as they get closer to the subway entrance, their walking pattern becomes more orderly until it appears to become a stream of human motion. As long as people stay on a straight path and no walkers come from the opposite direction (and no other obstructions appear), the rush-hour flow continues steadily and uniformly.

What happens when the parade turns the corner and descends into the subway? The flow slows and may even stop. People bunch at the stairway entrance and try to crowd in. The same bunching occurs on the platform when the train arrives and people crowd through its doors. But if there is an alternate route to the subway or if there are extra cars on the train, some of the people will walk in the direction of least resistance, and the pace of the flow will increase. Again when the car stops and the commuters get off, another congestion occurs that is relieved when the crowd eventually reaches a relatively open street.

Air flows through a ventilation system in much the same way. But resistance to air flow is not caused by stairs and other commuters. Air resistance in a ventilation system is caused by entrance into the hood, long runs of narrow ducts, sharp turns and changes in ductwork diameter, rough edges and surfaces inside the ventilation components, and discharge from the system.

Since your ventilation system may carry toxic materials, you want to be sure that air moves through it quickly and smoothly. Crowding, squeezing, and shoving will only slow air movement and possibly allow contaminated air to escape into the workspace air.

So in designing an economical and effective ventilation system, two primary principles are: plan for a minimum of air resistance, and select a proper fan. But that's not all. Below you will find other principles of good ventilation practice which you should consider as you plan your system and select equipment for it.

Rules for good ventilation

1. *Direct air flow away from the breathing zones of people who work in the area.* Likewise, do not position a worker in the path of contaminant emissions.

2. *Exhaust contaminated air from the workspace.* Do not blow it out. Pulling the air out of a workspace with an exhaust fan removes contaminants more effectively than attempts to blow them out, which often just disperse them around the work space.

3. *Place the exhaust opening of the ventilation system as close as possible to the source of the contaminants.*

4. *Avoid crossdrafts.* Open windows and doors, machinery, and human traffic can create air movement, which in turn can interrupt the flow of contaminated air towards the exhaust opening. Protect your operation from all crossdraft sources.

5. *Supply make-up air to replace the air exhausted by the ventilation system.*

If make-up air is not provided when large volumes of air are exhausted, the workspace will become starved for air and negative pressure will be created. Negative pressure in the workspace will increase resistance on the ventilation system and cause it to move less air. The negative pressure will cause air to move into the workspace through every available opening, leading to backdrafting of combustion and of other exhaust flues, to crossdrafts through windows and doors, and to pressure differentials. A useful indicator that negative pressure exists can be an outward-opening door. If the door is harder than usual to open, negative pressure very likely is the cause.

In many studios and shops, a simple air inlet, even a window, can supply sufficient make-up air. In other instances, fans and ductwork may be needed. The major cost associated with make-up air is heating it in the winter or cooling it in the summer, depending on where you live and the kind of work you do.

6. *Discharge the contaminated air away from openings that draw air into the studio or shop.* In too many ventilation systems, air exhausts are located very close to air intakes or to open windows. This may result in the reentry of contaminated air to the workspace. Such a problem may exist even in a home studio. Suppose a window fan removes hot soldering fumes from a basement studio. Another window fan on the floor above could pull the rising soldering fumes directly into a family's living quarters. No matter how simple your ventilation system, always check to see that removed air cannot recirculate into the building.

7. *Avoid polluting the community.* Toxic materials are problems for you in your workplace. So you can imagine the difficulties they pose for people outside. If toxins escape your workplace, they become potential hazards for your neighbors. People who live next to spray painters, for example, have been known to find themselves with speckled automobiles or houses if the wind is blowing the wrong way and if the spraying operation is vented improperly.

If your air outlet vents into a location with limited air flow (such as a courtyard or air shaft), you have a potential problem. Toxic materials can build up in areas like these, especially on hot, muggy summer days, when air hangs low, and when your neighbors have their windows open.

The extent of the environmental pollution problem depends primarily upon the toxicity, quantity, and concentration of material being used. Under no conditions should extremely toxic or corrosive materials such as carcinogens or strong acids be released into outside air. If you use these types of chemicals, you must have special air-filtration devices designed for your system.

Air levels for other, less toxic contaminants should meet the air quality levels set by local, state, and federal agencies. But remember that these current air pollution regulations reflect minimum standards. So if you, like most people, want to decrease air emission levels as much as possible, you should design a system capable of doing more than the legal minimum. Besides, it is possible that in the future the various governments will toughen their regulations. Anticipating stiffer future codes now could prevent expenses later.

Types of ventilation systems

There are two general classes of ventilation systems: *dilution ventilation* and *local exhaust ventilation*. Dilution ventilation does exactly what its name implies. By using a large volume of clean air, a dilution ventilation system dilutes a high concentration of a toxic material to an acceptable level, then removes the diluted mixture from the workplace. Although often cheap and easy to install, dilution ventilation has its limitations. It should be used only for removing solvent vapors or fumes of low toxicity or small amounts of moderately toxic vapors or fumes. Dilution ventilation never should be used to remove dusts, metal particles, or any highly toxic materials.

Where can dilution ventilation be used effectively? Common settings include darkrooms for the developing and printing of simple black-and-white photographs (no toning or special chemical processes), oil painting studios, and small lithography or offset print shops.

Local exhaust ventilation, on the other hand, has a wide range of applications. It is the major control method for removing materials of moderate to high toxicity, including gases, vapors, dusts, and fumes. And because it uses a system of hoods, ducts, air cleaners, and fans, a local exhaust ventilation system requires smaller amounts of air to remove contaminants than a dilution ventilation system does.

The major advantages and disadvantages of both dilution and local exhaust ventilation are described in Table 4. Review this information carefully. It should help you choose which system suits your particular situation. Once you have decided on either a dilution or local exhaust system, consult the detailed descriptions of system design in the following sections. They will help you decide on a properly functioning ventilation system for your needs.

Table 4

A Comparison of Dilution and Local Exhaust Ventilation

DILUTION VENTILATION

Advantages	Disadvantages
1. Low equipment and installation costs.	1. Does not eliminate exposure to contaminated air.
2. Effective control for small amounts of low and medium toxicity solvents.	2. Should not be used for high toxicity vapors and gases.
3. Effective control for flammable and combustible gases and vapors.	3. Should not be used for large amounts of any gases and vapors.
4. Requires little maintenance.	4. Ineffective for particulates (dusts, metal particles, metal fumes).
	5. Requires large volumes of heated or cooled make-up air.
	6. Not effective for handling surges of gases or vapors or irregular emissions.
	7. People working close to contaminant source can still have large exposures.

LOCAL EXHAUST VENTILATION

Advantages	Disadvantages
1. Captures contaminants at source and removes them from workplace.	1. System design and installation can be expensive.
2. Can handle all types of contaminants, including dusts, metal fumes, etc.	2. Requires regular cleaning, inspection and maintenance.
3. Requires small amounts of make-up air since uses low exhaust volumes.	
4. Low on-going energy costs because of low amounts of make-up air.	
5. Only alternative for high toxicity airborne materials.	

6 Dilution ventilation

Dilution ventilation mixes clean air with contaminated air. Thus it lowers the concentration of toxic substances to a safer level, then removes the entire volume of diluted air from the workspace. A dilution ventilation system has two major components: a source of make-up air (an intake fan or some form of natural replacement), and an exhaust fan. The exhaust fan should be located so that it pulls clean air through the breathing zones of people in the workspace. When designing a dilution system, you should satisfy the requirements of two specifications: (1) How much exhaust volume is required to dilute the contaminants so they are safe? And (2) what fan is appropriate to control the flow of air through the work area?

Exhaust volume

If you decide that dilution ventilation will provide the "adequate ventilation" called for on the labels of the products you use, you must determine what volume of air must be moved over a set period of time to dilute the contaminant to a safe level. This volume of air per unit of time is referred to as the exhaust volume or the air capacity, and it is measured in cubic feet of air per minute (cfm).

The procedure for determining the exhaust volume is described below.

1. Find in Table 5 the required air volume for diluting the total amount of solvent to the TLV.

2. Multiply the air volume calculated above by a multipurpose safety factor, K. You yourself will determine this K value without the aid of a table or formula. The K value is used to maintain the concentration of the contaminant well below the TLV. K values vary from three to ten, depending on several factors:

The toxicity of the solvent.
Slightly toxic material: TLV equal to or more than 500 ppm
Moderately toxic: TLV from 100 to 500 ppm
Highly toxic:* TLV equal to or less than 100 ppm

*Use dilution ventilation only with small amounts of solvents with TLVs less than 100 ppm.

Rate of evaporation of the solvent.
A high evaporation rate requires a higher K.

How close you are to the solvent evaporation.
Are you doing fine, detail work where you are very near the solvent vapors, or are you separated from the solvent by the length of your arm, a work table, or several feet?

How effectively the vapors are cleared from the work area.
Is air flow away from your breathing zone? Are there corners or areas in the studio where air doesn't circulate and vapors can accumulate?

Review your process and your space and use good common sense to select the proper K value. Remember, the higher the K value, the more protection is provided. Erring on the side of caution will protect your health.

3. Determine the required air capacity over the set period of time by dividing the air volume calculated in step 2 by the number of minutes the solvent is used. Remember to allow time for evaporation of the solvent.

If the solvent you use does not appear here, you will have to make additional calculations using this formula:

$$\text{Cubic feet of air per pint evaporated} = \frac{403 \times \text{sp. gr. liquid} \times 1{,}000{,}000 \times K}{\text{M.W. liquid} \times \text{TLV}}$$

where

sp. gr. = specific gravity of the liquid
M.W. = molecular weight of the liquid*

Then use Step 3 to determine air capacity required per unit of time.

Let us consider an example of how to do this calculation. Assume that a printmaker uses one cup (a half pint) of turpentine to clean a press bed. Assume that this operation and the time for the solvent to evaporate is sixty minutes.

From Table 5, you can see that 25,500 cubic feet of air are required to dilute each pint of turpentine to the TLV. For a half pint, you need 12,750 cubic feet of air. Because turpentine evaporates relatively quickly, and because it is borderline highly toxic (TLV = 100), and the printmaker is working close to it, a K value of ten should be chosen. Therefore, to dilute a half pint of turpentine to a safe level, multiply the 12,750 by ten to give 127,500 cubic feet of air.

The required air capacity then becomes 127,500 cubic feet divided by sixty minutes, which equals 2,125 cubic feet per minute (cfm).

This means that a fan will be required that can exhaust 2,125 cfm. Note that this is a large volume of air. If the room is small then you will need to provide make-up air, which could make the room chilly in the winter unless it is heated.

* Source: ACGIH Committee on Industrial Ventilation: *Industrial Ventilation: A Manual of Recommended Practice,* 17th ed. (Michigan: ACGIH, 1982).

A word of caution: The illustration involves a single solvent. Usually several solvents or mixtures are used at one time. The cumulative toxic effect is difficult to predict. The best advice is to consider the overall health effect as additive, meaning that you should assign each solvent a percentage according to the amount used. So calculate the exhaust volume of air required for each solvent separately, and then add these volumes to determine the total amount of air to be exhausted.

Table 5

Dilution Air Volumes For Solvent Vapors

Liquid	1982 TLV	Air volume (standard air) for dilution to TLV (cubic feet/pint evaporated)
Acetone	750	7,300
n-Amyl acetate	100	27,200
n-Butanol (butyl alcohol)	50*	88,000
n-Butyl acetate	150	20,400
Cyclohexane	300	12,400
Ethyl acetate	400	10,300
Ethyl alcohol	1,000	6,900
Isoamyl alcohol	100	37,200
Isopropyl alcohol	400	13,200
n-Heptane	400	6,900
n-Hexane	50*	61,700
Kerosene	100	30,000–35,000
Methyl acetate	200	25,000
Methyl alcohol	200	49,100
Methyl chloroform (1,1,1-trichloroethane)	350	11,400
Methylene chloride	100	63,400
Methyl ethyl ketone	200	22,500
Methyl isobutyl ketone	50*	64,600
Methyl propyl ketone	200	19,000
Mineral spirits (Stoddard solvent)	100	30,000–35,000
Naphtha, VM&P	300	9,000–11,000
n-Propyl acetate	200	17,500
Styrene mononer	50*	69,900
Toluol (toluene)	100	38,000
Turpentine	100	25,500
Xylol (xylene)	100	33,000

* Use dilution ventilation only with small amounts of solvents with TLVs less than 100 ppm.

Control of air flow

Successful dilution ventilation depends not only on the correct exhaust volume, but also on the control of air flow through the workspace. Careful positioning of air inlets and outlets will assure proper control. Remember:

1. Move only clean air through a worker's breathing zone.
2. Position the workspace to avoid crossdrafts.
3. Locate the workspace and air inlets and outlets so that other workers in surrounding areas are not exposed inadvertently.

Figure 4 Good and Bad Dilution Ventilation

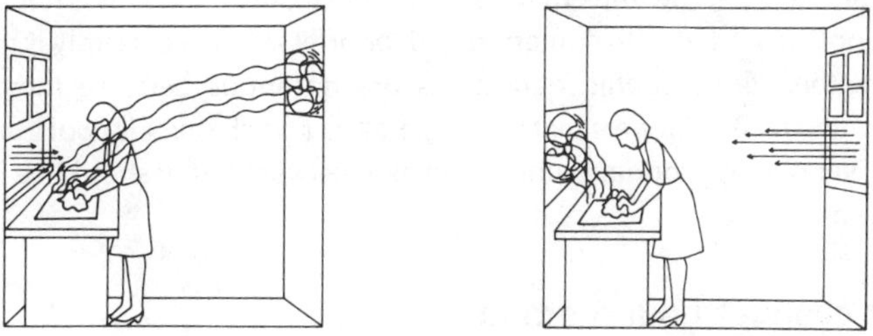

BAD GOOD

Adapting heating and air conditioning ventilation systems

Mechanical ventilation systems are designed to provide comfortably heated or cooled air. Many of these systems are known as closed systems because they recirculate a high percentage of the air involved in order to reduce energy costs for heating or cooling. Typically, a closed system does not have windows that open. Consequently, the amount of outside air entering the space is minimal. Many schools and office buildings have these closed systems. Similarly, room air conditioners recirculate almost all of the room air, even in vent mode.

Such systems, however, are not designed to removed contaminated air. Even so, a ventilation specialist sometimes can make design changes to provide dilution ventilation. If you have a system that can be adapted, you must observe the following steps to assure that the proper adjustments are made:

1. Determine the required exhaust volume for safe dilution of the substance used. Increase air flow into the room from existing air inlets by this amount, or provide a separate make-up air system.
2. Close off existing air return registers to prevent contaminated air from circulating through the building. Do not just shut them. Seal them thoroughly and permanently so there are no leaks.
3. Install new air outlets which exhaust directly to the outside atmosphere. Avoid exhausting air near intakes of any buildings.

4. Have the system rebalanced by a qualified ventilation specialist to be sure that all rooms in the building receive adequate incoming air. The increase in air flow for dilution ventilation may lead to a decreased volume of air coming into other parts of the building.

A final word of caution about dilution ventilation: *do not use standard room air conditioners for exhausting contaminated air or for bringing in fresh air!* They do not have outside air-intakes capable of the large amounts of air that are needed by dilution systems. They recirculate most of the air in a room, so they cause buildup, not a dilution, of contaminated air. Air conditioner filters are intended only to trap large dust particles that can clog a conditioner's cooling element. They do not filter harmful contaminants such as gases, vapors, or fine dusts.

Air conditioners deceive persons in their vicinity because they circulate cool, satisfying air. Most people think that cool air is clean, fresh air. This is not necessarily the case. Cool air loaded with contaminated, odorless vapors certainly is not clean or safe to breathe. Photographic darkrooms are notorious because they often are equipped with air conditioners that keep operators and chemicals cool but do not get rid of toxic vapors. Appropriate supply and exhaust ventilation should be installed in all darkrooms.

Room air changes/hour method

Many engineers will express dilution ventilation needs in terms of a certain number of room air changes per hour. For example, Kodak recommends a minimum of ten room air changes/hour for ventilation of black-and-white darkrooms.

An exchange rate of ten air changes/hour means that the air in the room will be completely replaced ten times an hour, or every six minutes. You can use the following formula to convert from air changes/hour to cubic feet per minute (CFM):

$$\text{CFM} = \frac{\text{(room volume in cubic feet)} \times \text{no. of air changes/hour}}{60}$$

There are two major problems associated with this method. First, it provides the same amount of ventilation for a room regardless of the amount of gas or vapor that must be exhausted. From the previous discussion, you know that larger amounts of gases or vapors require larger amounts of dilution ventilation. Second, it provides less ventilation for a small room than a large room—even though the same amount of gas or vapor would be more concentrated in the smaller room.

What are the solutions? Use the air changes/hour method only if there is no other way to calculate the needed amount of dilution ventilation. And for smaller rooms or if you are using larger amounts of chemicals than usual, increase the number of air changes/hour. Small darkrooms, for example, should perhaps have twenty instead of ten room air changes/hour.

7 Local exhaust ventilation

Local exhaust ventilation removes airborne contaminants at their source. If a local exhaust system is designed properly and works effectively, contaminated air never will be seen, smelled, or breathed in the studio or shop it serves. Processes that use highly toxic materials or large amounts of toxic materials, and processes that produce dusts or other solid particles should be controlled by local exhaust ventilation, not by dilution ventilation (see Table 6).

There are four major components in a local exhaust ventilation system: a hood, ducts, an air cleaner, and a fan. Contaminants are drawn in through the hood, transported inside the ducts, removed by an air cleaner (if used), and moved

Table 6

Examples of Processes and Materials Requiring Local Exhaust Ventilation

Abrasive ("sand") blasting	Metal foundry furnaces
Acid etching	Metal melting and casting
Acid pickle baths	Mixing dry clays
Aerosol sprays	Photoetching
Air brush	Photographic toning
Burn-out kilns	Photolithography
Color photographic processing	Plastic resin casting
Electroplating	Powered stone-carving tools
Furniture stripping and refinishing	Screen print drying
Glassblowing furnaces	Screen printing and cleaning
Glaze spraying	Soldering
Grinding	Spray operations
Heated dye baths	Welding
Hydrofluoric acid etching	Woodworking shops
Kiln firing (gas and electric)	

the system by a fan. A local exhaust ventilation system *always* needs a y of replacement air. This may or may not require a fan to supply the air.

Good planning will result in a system that is efficient and economical to operate. Planning begins on paper, not with a hammer and crowbar. Always lay out a system on paper first. Adjustments and corrections are easily made on paper, but they can be costly and time-consuming when perfomed on the actual components. An effective local exhaust ventilation system must be designed carefully and the proper components selected with equal care.

Factors in design

These factors must be considered:

1. *Air Capacity.* The first specification to be determined is air capacity (Q), also called air quantity. It is the volume of air moving per unit of time. In calculating the air capacity of a local exhaust system, the first considerations are to identify the processes it is supposed to handle and the types of contaminants generated. Then you can choose an air capacity and the design components (hood and ductwork) that will maintain that air capacity through the system.

Air capacity is computed by using the formula $Q = AV$, where:

Q = volume of air flow in cubic feet per minute (cfm)

A = cross-sectional area that the air flows through in square feet (ft^2)

V = velocity of air flow in feet per minute (fpm).

2. *Air Pressure.* As described at the beginning, differences in air pressure cause air to move from one location to another. In ventilation, air pressure is measured as the difference between the actual pressure and atmospheric pressure. It is measured in inches of water—that is, how high the difference in pressure can lift a column of water. Pressure is negative if below atmospheric pressure and positive if above atmospheric pressure.

There are three types of pressure: static pressure, velocity pressure, and total pressure. Static pressure is the potential pressure exerted in all directions by air at rest. It has been described as the tendency of the air to cause its container to burst or collapse, and is either positive or negative. The static pressure of air in a duct on the suction side of a fan is always negative.

Velocity pressure is the pressure needed to make air move at a given velocity. It is always positive. For standard air, with atmospheric pressure at sea level and temperature at 70° F. (20°C.), the velocity pressure is related to the air velocity by the equation:

$$v = 4005 \sqrt{VP}$$

where v = velocity, and VP = velocity pressure.

Table 14 in Appendix B lists the values of the velocity pressure at various air velocities. Note that these are for standard air. If your situation differs considerably from standard air—for instance, if you work at an altitude greater than 2,000 feet—corrections will have to be made using the manual listed previously, *Industrial Ventilation: A Manual of Recommended Practice*.

The total pressure is the algebraic sum of the static pressure and the velocity pressure.

3. *Air Resistance*. The amount of air resistance determines the specifications for your system's fan. Air resistance, or the total resistance of a local exhaust ventilation system, is the sum of the resistances exerted by the hood entry, duct friction, air cleaner resistance, and discharge resistance.

Hood entry losses are caused by turbulence and the acceleration of the air entering the hood.

Duct friction is caused by air rubbing against itself and the interior walls of the ductwork, and is influenced by changes in air flow direction and duct diameter.

Fabric air cleaner resistance occurs when air moves through a filter. As the filter gets dirtier, resistance increases.

Discharge resistance occurs when the exhausted air slows down as it meets the outside atmosphere.

Air resistance is usually expressed as static pressure in inches of water. When you design a ventilation system, calculate overall resistance before selecting a fan. See Appendix B for computing a system's total resistance.

4. *Hoods*. Hoods are complete or partial enclosures which draw contaminated air into a ventilation system. A good hood design uses the smallest possible air volume to capture and remove a contaminant. Therefore, the more a work process can be enclosed, the smaller the required air capacity will be, and the more effective will be the hood.

8 Hoods for local exhaust ventilation systems

You will find that many types of hood designs are available. This handbook describes the types used most commonly by artists, craftspeople, and schools. They are: spray booth, slot hood, canopy hood, movable exhaust system, and built-in hoods for dust-producing machines. Some hoods, such as spray booths, and some types of movable exhaust systems are available commercially. You can purchase these directly through a manufacturer's catalog. You may, however, have to purchase ducting separately. Other hoods, such as canopy and slot hoods, must be designed and fabricated individually.

Your selection of an appropriate hood depends on several factors:

Toxicity of the material to be captured
Form of the contaminant (dust, fume, gas vapor)
Manner of release: how the contaminant is released into the air, continuously or in sudden surges

A comparison of the advantages and disadvantages of each hood type together with a list of some of the processes for which each type can be used are presented in Table 7. Before you read the next section, review this table carefully to determine which hood type may be appropriate.

General rules for hood design

Each type of hood has specific design requirements, but a few general principles apply to all hoods. If you follow these principles, they will increase the effectiveness of your system.

1. Eliminate air currents around the work process before installing an exhaust hood. Air flow from open windows, doors, cooling fans, human traffic, and

Table 7

Local Exhaust Ventilation for Art and Craft Processes

CANOPY HOOD *Process:* ceramic kiln, metal foundry, forging. *Form:* vapor, gas, fume, steam. Not ready made.*	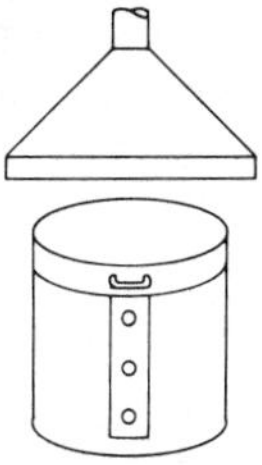	**Advantages** Captures upward-moving contaminants. Good for heat-producing operations. Handles surges of hot air. Easy to construct. Designed in various sizes. **Disadvantages** Exposes worker to contaminants if he works under hood. Often used improperly. Should not be used for dust.
SLOT HOOD *Process;* silkscreening, acid etching, bench welding, soldering. *Form:* vapor, gas, fume. Not ready made.*	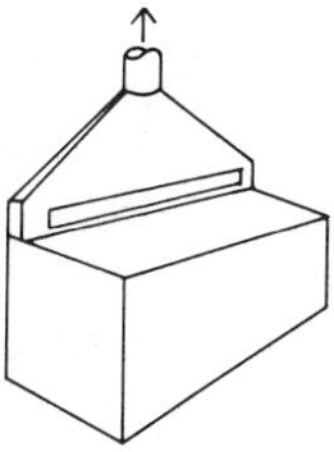	**Advantages** Has universal applications. Does not interfere with most processes. Can cover a long work area: four-foot maximum selections are used side by side. Good for processes performed on a work bench. **Disadvantages** Work must be kept in close proximity of slot opening. Work-table depth has a 36″ maximum. Can be difficult and costly to construct.
SPRAY BOOTH and other enclosed hoods *Process:* spraying laquer, paint, ceramic glazes, flammable materials, highly toxic materials. *Form:* vapor, gas, fume, dust. Ready made.*	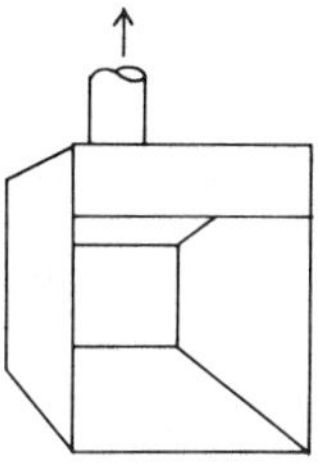	**Advantages** Completely encloses contaminant. Safest hood for flammables and highly toxic materials. Conserves materials. Reduces housekeeping. Designed in various sizes. **Disadvantages** Cost of operation high. Requires more space than other hoods. Requires constant cleaning.
PLAIN-OPENING HOOD *Process:* welding, soldering, art conservation. *Form:* vapor, gas fume. Ready made.*	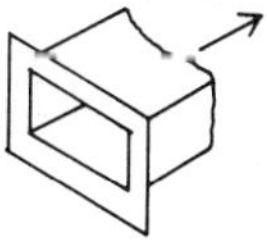	**Advantages** Can be used in areas where conventional hood will not fit. Can be connected to flexible ducting for repositioning to fit work and for several alternating locations. Easy and inexpensive to construct. **Disadvantages** Must be in a range of four to six inches from opening for effective capture.
DUST-COLLECTING HOOD *Process:* grinding, woodworking, polishing. *Form:* dust. Sometimes ready made.*	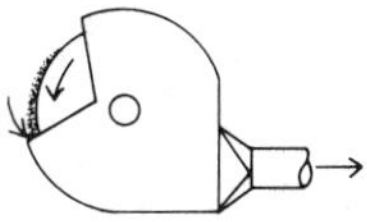	**Advantages** Reduces housekeeping. Reduces fire hazards. **Disadvantages** Ready-made hoods often ineffective. Requires constant cleaning.

* Ready-made hoods are available in set sizes and are complete with all equipment from manufacturer and dealers. Others must be fabricated by a sheet-metal shop.

heating units may create crossdrafts which can interfere with effective capture of the contaminated air.

2. Minimize resistance through good design. Air drawn into a hood encounters resistance, called hood static pressure. Hood static pressure is the sum of the hood entry loss and an acceleration loss. As air is drawn into the hood, it is squeezed together tightly and moved into the hood at an accelerated rate. (Remember the commuters and the subway entrance.) The hood entry loss has been calculated for some types of hood, and manufacturers generally will furnish the hood entry loss figures for the hoods they make, such as spray booths and movable exhaust systems. But if you do not know the hood entry loss, you can calculate it yourself. Refer to Appendix B.

3. Enclose the hood as much as possible. For example, a booth provides a more efficient enclosure than a slot hood. Use side curtains and flanges (a flat rim around the edge of a plain, open hood) or baffles to reduce the amount of air exhausted. These can reduce the required air capacity by 20 percent to 30 percent. When possible, position hoods on tables and against walls, which act as partial enclosures. For example, a slot hood placed on a tabletop uses the table surface as a baffle (Figure 5); a canopy hood placed against a wall makes use of the wall itself as a convenient enclosure.

4. Position the hood as closely as possible to the source of the contaminants. The farther they are from the hood, the greater the air capacity required to remove the contaminants. (See Figure 6.)

5. Keep contaminated air away from the breathing zones of the operator and other people in the workspace.

6. Furnish a system that provides a capture velocity which draws the contaminants completely into the hood. Use the recommended minimum capture velocities listed in Table 8 as guidelines.

Figure 5 Enclosing the Hood

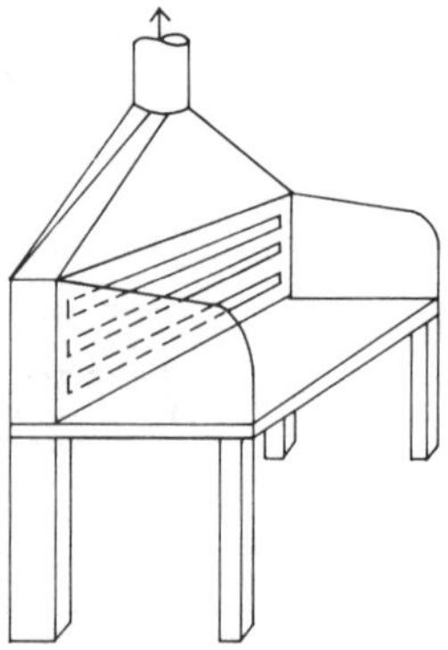

Table 8

Capture Velocities

Release of contaminant	Examples	Capture velocity, fpm
No velocity into still air	Evaporation from print drying	50–100
Low velocity into moderately still air	Spray booths, welding, plating, etching	100–200
Moderate velocity into rapidly moving air	Kilns, clay mixing, foundry furnaces	200–500
High velocity into very rapidly moving air	Grinding, abrasive blasting, woodworking	500–2000

7. Take advantage of natural movement of contaminants into the hood. For example, heat rises. So place a canopy hood above a kiln to trap the upward moving gases and heat. Or equip a grinding wheel or woodworking machine with a partial enclosure to trap particles as they spin off.

8. Paint iron elements of a system to retard corrosion. Fabricated hoods normally are made of welded 16-gauge black iron, which can be suitable if the contaminants are not corrosive. But black iron eventually will corrode when exposed to acid gases. Painting black iron both inside and outside may slow the corrosion. You also may wish to use more expensive stainless steel.

Figure 6 Placement of Hood

Place hood as close to the source of contaminant as possible.

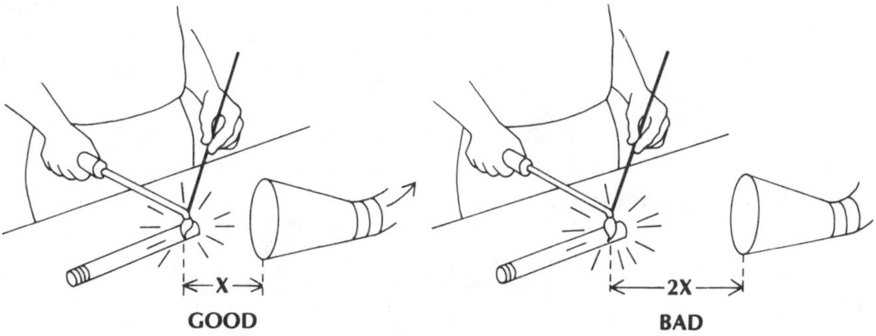

If 1000 cfm is needed at x distance, 2x requires 4000 cfm.

Source: Committee on Industrial Ventilation, American Conference of Governmental Industrial Hygenists. Adapted from Figure 4-9 of *Industrial Ventilation: A Manual of Recommended Practice,* 17th ed. (Michigan: ACGIH, 1982)

Canopy hood

Most canopy hoods are installed permanently above a specific spot. In a few cases they can be equipped with swinging joints so that they can be moved out of the way. The swinging joint may, for example, allow easy access to top-loading kilns. In the case of some small kilns, the hood can be stationary and the kiln itself can be moved.

Fireproof side curtains installed on two or three sides of a hood to enclose a process will greatly increase a hood's efficiency. Curtains reduce the amount of non-contaminated air which a hood pulls in, they decrease the effect of crossdrafts, and they further protect an operator and others working in the area. Side curtains are made of flexible strips, sheet metal, or woven fiberglass cloth, depending upon the specific operation and temperature.

A canopy hood should overlap the sides of a contaminant's source and should be positioned as close to the source as possible without interfering with the work. An optimum distance is not more than two feet above the work process, but be sure to allow for head room so that you do not bump into the hood. A common included angle for canopy hoods is ninety degrees. (See Figure 7).

To remove most contaminants adequately with a canopy hood, you should provide an air capacity of 100 cfm for every square foot of hood face. This figure is based upon experience. Other ways to calculate air flow requirements for canopy hoods are discussed in the ACGIH manual on industrial ventilation. Refer to Appendix B in the manual for a specific example of how to apply these design requirements.

Figure 7 Canopy Hood

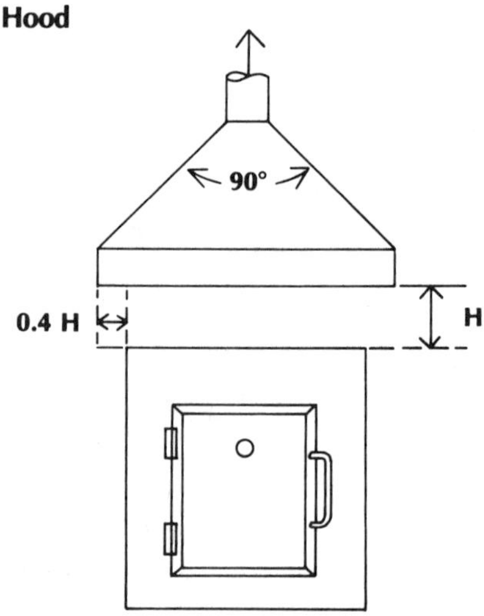

Spray booth

Paint spray booths are available from commercial suppliers, directly from manufacturers, and at used-equipment outlets in a variety of models. Most artists and craftspeople find either a table model or a floor model suitable, depending on how they use it. Use a table model for small, easy-to-transport items. Use a floor booth or even a spraying room for very large items or for objects that must be wheeled around. Always place a booth where there is sufficient space to move pieces in and out.

Design a booth to handle your usual maximum work piece. The actual working depth inside a booth is smaller than its outside depth dimension because a booth requires a filter or baffle at its rear, and this device takes up space. The working depth should accommodate the piece *inside* the booth, not at its edge, and it should provide some clearance at the booth's rear. Unless otherwise recommended by the manufacturer, booth dimensions (width, height, and depth) should be ten percent larger than the usual maximum piece.

Spray booths are equipped with some type of air-cleaning device to trap overspray and paint particles. The primary types are baffles, filters, and water washes. These air cleaners control the number of particles going through a system and help to protect your neighbors' property from discharges. Most artists and craftspeople find that filters work well for light and intermittent use. The filters are made from thick, blanket-like material and are installed at the rear of a booth. Filters must be cleaned and changed regularly to minimize resistance in an exhaust system.

Figure 8 Spray Booth

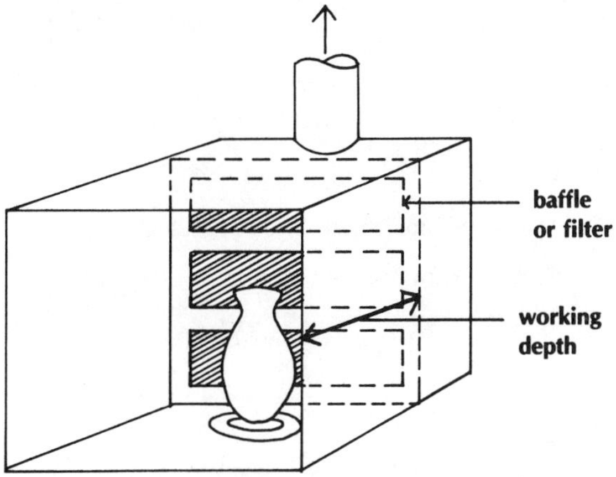

Unless purchased second-hand, most spray booths come equipped with a propeller fan and a short length of connecting ductwork, which is acceptable when there is only a short connection duct before discharging outside. With longer runs of ductwork, always install the fan at the end of the duct. If the fan is built into the booth or immediately inside the duct, rather than at the end, solvent vapors can leak through holes in the ductwork because of the positive pressure created just beyond the fan.

Spray booths generally come equipped with either a back or top discharge. Choose the one that will make connecting the duct easier.

Recommended face velocity (V) for spray booths are:*

Minimum:	
Light-to-medium production	125 fpm
Average:	
Heavy production	150 fpm
Maximum:	
Heavy particles (like ceramics)	200 fpm

Maintain spray booths regularly and clean a booth's walls of all paint buildup to eliminate fire hazards. To make cleanup easier, place flame-retardant paper over the inside walls, then remove it when it becomes covered with paint. Change filters regularly. Inspect fan blades and ducts for paint residue and clean them when necessary.

Spray booth: preventing fire and explosion

Government agencies have issued strict standards to prevent fire and explosion in spray booths. Flammable vapors can reach explosive levels in the enclosed space of a spray booth, so it is critical for you to observe very carefully the requirements set by OSHA or other agencies for your operation. Here are some of the rules for safe spray booth use. They are taken from OSHA standard 1910.107 for spray finishing using flammable and combustible materials. (See the standard itself for complete rules. Also see NFPA 33 Spray Applications, which is updated regularly.)

Most local codes require a sprinkler head in spray booths.

A booth shall have a minimum of three feet of clear space around it.

No open flame or other source of ignition shall be within ten feet of a spray booth opening. All electrical equipment within this area must meet Class I Division 2 requirements (explosion-proof).

*These are ACGIH recommendations. OSHA standards require a minimum face velocity of 100 fpm.

The discharge shall be at least six feet from any combustible outside wall or roof.

Booths shall have smooth, easy-to-clean inside surfaces so that residues of sprayed materials cannot accumulate.

Spraying areas shall be kept clean of combustible residues. All tools and scrapers shall be made of non-sparking material.

All light fixtures within twenty feet of a spray booth shall be totally enclosed and shatter-proof.

All metal parts which transport flammables in a ventilation system and all electrical tools shall be properly grounded.

A fan shall be made of non-ferrous or non-sparking material.

Electric motors which drive a fan shall not be placed inside a booth or ductwork. Drive belts should be grounded to prevent static buildup.

Solvents used for cleaning shall not have a flash point less than 100°F.

Smoking is prohibited and "no smoking" signs shall be posted in all spraying areas.

The ventilation system shall be operated during spraying, immediately after spraying to remove all vapors, and during drying time of the sprayed articles.

Average velocity across the face of a booth opening shall be a minimum of 100 feet per minute.

A gauge or other type of warning device shall be installed across the filter to indicate a drop in the air velocity through the filter. The filter shall be replaced immediately to maintain velocity.

Used filters shall be disposed of properly and immediately.

Slot hood

A slot hood has an exhaust opening with a width less than one-fifth of its length. The exact dimensions depend upon the length of the work area to be controlled and the required air capacity (Q). However, maximum slot length is generally four feet for ease of exhausting and because material for longer ones is more difficult to obtain. If you need a slot opening longer than four feet, you should use an additional hood. A plenum (air pressure equalizing chamber) is placed behind the slots to slow the air stream as it is pulled into the hood through the slots.

If possible, equip a slot hood with side shields or curtains, place it on top of a work table so the tabletop acts as a bottom flange, or do both. These measures partially enclose the process and reduce by as much as 25 percent the amount of air handled.

Most slot hoods have an effective capture depth of no more than thirty-six inches. Contaminants generated at any greater distance will not be completely captured and pulled into the hood. This distance requirement can pose a problem if the slot hood is placed on a very deep work table. To compensate for extra depth of the work surface, keep your work as close to the hood as you can.

Recommended velocities for a slot hood are as follows:

Slot velocity	1,000–2,000 fpm
Plenum velocity	500–1,000 fpm (should be half of slot velocity)
Duct velocity	1,000–4,000 fpm (for economy 2,000 fpm is often used)

For a slot hood with side shields, air capacity (Q) of 350 cfm per linear foot of work area should be sufficient for most operations. (This is an empirical figure and should be revised depending on the process.)

Figure 9 Slot Hood

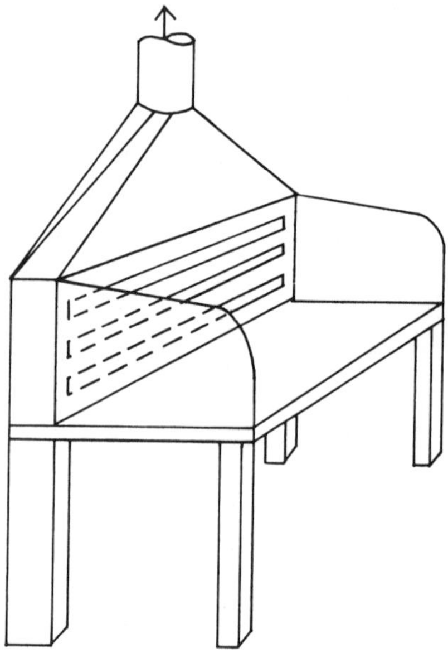

Figure 10 Plain Opening Hoods

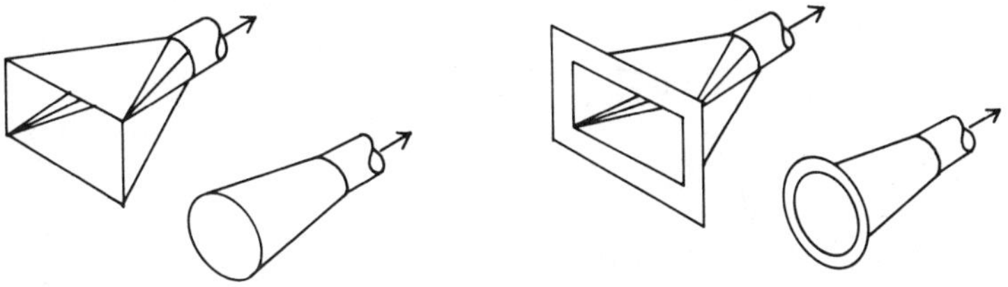

Plain opening hood

A plain opening hood is the simplest of all hood designs. It is merely the open end of a duct. (See Figure 10.) To make the hood effective, place the opening as close as possible to the source of contamination, because the velocity of air directed towards the hood opening is inversely proportional to the square of the distance from the hood to the source of contamination. The efficiency of a hood can be improved greatly by placing a flange or flat rim around its opening to direct the flow of air into the opening. Plain opening hoods in the shape of a tapered cone offer greater capture efficiency, and they are available commercially.

Plain opening hoods with flexible ducting are common in welding operations where the ability to move the hood around is crucial. Commercial portable ventilation systems intended for welding are readily available. Plain opening hoods with flexible ducting also can be used to capture solvent vapors and other contaminants in situations where mobility is needed or where a fixed exhaust system is not practical. Examples include painting restoration, set design, and sculpting.

With plain opening hoods, the crucial question is not the face velocity at the hood opening, but the capture velocity at the source of the contamination. The air velocity at the source must be strong enough to pull the contaminants into the plain opening. For solvent vapors and gases, the capture velocity is normally set around 100 fpm; for welding fumes, around 200 fpm. (Also see Table 8.)

The air capacity for the plain opening hood can be calculated from the equation $Q = V\ (10X^2 + A)$

where Q = air capacity, in cfm

V = capture velocity, in fpm

X = distance from source to hood opening, in feet

A = cross-sectional area of hood opening, in square feet.

Normally the cross-sectional area A is much smaller than $10X^2$, and can be ignored. Therefore the equation for air capacity becomes $Q = V\ (10X^2)$.

Figure 11 Grinder Hood

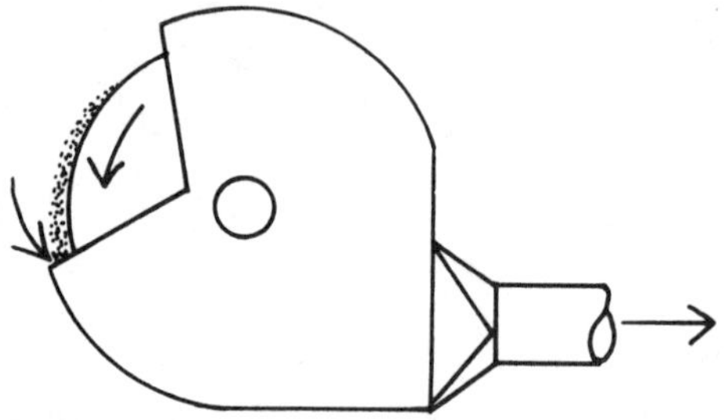

Dust collecting hoods

Hood design for a dust-producing machine follows the same principles applied to the other systems. The dust-producing part should be enclosed as much as possible without hindering the work process, and the hood design should take advantage of the natural movement of particles.

Because transporting dust and other particulate matter through a vented system can cause some problems, special care must be taken to avoid clogging the ductwork and allowing material to accumulate on the fan blades and other components. This is extremely important if flammable materials, such as wood dusts, are handled. An air cleaner, such as a cyclone or fabric collector, usually is employed to collect the dust. Also, the ductwork should be equipped with access openings, especially near its elbows, to enable removal of accumulated material.

Air capacity will vary according to the size and type of exhaust hood. A minimum duct velocity of 3,500 fpm usually is recommended for dusts.

9 Ducts for local exhaust ventilation systems

Once a hood captures the contaminated air, it is carried away through ducts and is discharged. Because the ductwork composes a large portion of a ventilation system's resistance, you should plan duct size and layout carefully in order to minimize resistance.

In addition to the analogy of pedestrian traffic, you can compare the flow of air to the flow of water to visualize what happens when the flow meets resistance. Imagine a garden hose. Water flows through it evenly and smoothly when the hose lies straight along a smooth lawn. However, as soon as the hose is kinked up, knotted, or flattened, the flow may be reduced to a trickle or stopped completely. The same thing happens to air when it is confined in a local exhaust ventilation system if the ductwork is not designed and constructed properly.

Rules for duct design

Follow these simple rules to keep air flowing smoothly, the contaminated air inside the local exhaust system where it belongs, and the costs down.

1. *Use round ductwork wherever possible.* Rectangular ductwork should be used only with caution because it has greater air resistance and it can accumulate dust. Ducts' interior surfaces should be clean and smooth. Air passing over smooth surfaces creates less friction and therefore meets less resistance. For normal purposes, use galvanized sheet steel. Rigid plastic duct can be used in systems which handle highly corrosive materials provided the materials do not attack the plastic. Plastic ducts are usually very expensive.

2. *If possible, keep ductwork short.* The longer the duct run, the more resistance the air encounters. If you can, always conduct your work process close to the discharge outlet, so you can minimize duct length. Keep in mind that each foot of duct adds to a system's cost.

3. *Use recommended duct velocity for a specific process.* For any given air capacity (Q) in the duct, the velocity varies with different duct diameters (Q = AV). Select a duct size that will exhaust air from the hood at the economical velocity of 2,000 feet per minute (fpm) for solvents and gases. For normal dust-carrying systems, the duct velocity should be at least 3,500 fpm. Refer to Table 9 for descriptions of airborne contaminants and recommended minimum design (duct) velocities.

4. *Minimize turns and size changes.* Each change in direction or change of duct size increases resistance. When turns or duct diameter changes are necessary, keep them soft and gradual.

5. *Balance compound systems.* If more than one hood is connected to the same ductwork and fan, balance the air flow through each branch. For some types of systems, a ventilation engineer will have to do the balancing. Balanced air flow can be designed as a permanent feature of a system by using specific duct diameters and lengths to deliver the required air capacity. Alternatively, dampers (also called blast gates) can be installed inside the duct branches, and these dampers can be adjusted to accommodate the required air capacity for each hood. Dampers make a system more flexible, but may cause it to malfunction if they are improperly adjusted. Systems that exhaust toxic substances should be balanced and certified by professional engineers.

6. *Eliminate leaks and holes in the ducts.* Contaminated air can leak into a work area if a system is under positive pressure. This is a good reason to install the fan at the discharge end, keeping the ductwork under negative pressure.

7. *Inspect and clean ducts regularly.* Clogged air ducts restrict air flow and increase resistance. If you use flammable materials, do not let them build up in the ducts because they can be a fire hazard.

8. *Extend the discharge section of ductwork at least two duct diameters from the fan.* This will keep the fan running at or near its potential.

Duct size

Duct size depends upon the air velocity needed in the duct. One general principle is that the smaller the duct size for a given air capacity, the greater the velocity of air inside the duct and the greater the frictional resistance in the system.

Sheet metal workers can construct ducts to practically any size desired. However, certain standard sizes are readily available and are therefore cheaper. Round ducts of four inches to twelve inches in diameter are made in one-inch increments and usually in half-inch increments. For sizes between twelve inches and eighteen inches, only one-inch increments are usually available. For diameters greater than eighteen inches, increments vary from one to three inches.

Ducts also come in various thicknesses or gauges. In general, for ducts up to eighteen inches in diameter, twenty-gauge galvanized steel is used. For between eighteen inches and thirty inches, eighteen-gauge is generally used. Most elbows and transitions are generally made from sheet metal two gauges thicker than the duct. Most hoods are made of sixteen-gauge galvanized or black steel. For more details, consult local codes and *Industrial Ventilation: A Manual of Recommended Practice*.

Table 9

Range of Duct Design Velocities

Nature of contaminant	Examples	Design velocity
Vapors, gases, smoke	All vapors, gases, and smokes	Any desired velocity (economic optimum velocity usually 1200–2500)
Fumes	Welding, soldering, & foundry fumes	1400–2500
Very fine light dust	Cotton lint, wood flour, litho powder	2000–3000
Dry dusts and powders	Fine rubber dust, jute lint, cotton dust, shavings (light), leather shavings	2500–3500
Average industrial dust	Sawdust (heavy and wet), grinding dust, buffing lint (dry), shoe dust, granite dust, silica flour, general material handling, brick cutting, clay dust, foundry (general), limestone dust	3500–4000
Heavy dusts	Metal turnings, foundry tumbling barrels and shakeout, sand blast dust, brass turnings, cast iron boring dust, lead dust	4000–4500
Heavy or moist dusts	Lead dust with small chips, moist cement dust, asbestos chunks from transite pipe cutting machines, buffing lint (sticky), quicklime dust	4500–5500

Source: Committee on Industrial Ventilation, American Conference of Governmental Industrial Hygenists. Adapted from Table 4-2 of *Industrial Ventilation: A Manual of Recommended Practice*, 17th ed. (Michigan: ACGIH, 1982).

Flexible ducts

Flexible ducting or hose, used for movable exhaust systems, deserves special mention. Flexible ducting is spiral in shape and usually is made of fabric or metal. Fabric hose consists of a fabric (usually cotton with a neoprene coating) over metal wire. Flexible ducts generally are used for normal temperature and non-corrosive applications—for example, as a dryer hose. You may need metal hose for high temperatures, corrosive operations, some solvents, or dust. Metal hose usually is made of aluminum or stainless steel.

Flexible hose is readily available in one-inch increments for diameters from four inches to eight inches, and in two-inch increments from eight inches to twelve inches. Smaller sizes also are available.

Flexible hose resistance usually is much higher than the resistance of conventional galvanized ducts. With flexible ducts, resistance is further increased by kinks that occur when the hose is not extended to its full length. To decrease this resistance: (1) use the least amount of flexible hose possible; (2) eliminate any sharp 90-degree bends in the system; and, (3) cut off any excess hose. As a general rule, if you double the resistance figure for a rigid duct, you will have a rough estimate of the resistance of a flexible duct system.

Flexible ducting's principal advantage is the maneuverability of the plain opening hood at the end. Locate the hood as close as possible to the source of contamination.

Flexible ducts are maneuvered in two ways. The first involves mechanical arms similar to those used to control movable desk lamps. These arms can be purchased commercially. The second method involves two or more overhead pulleys with a rope attached at one end to the duct and at the other to a cleat device.

The commercial mechanical arm devices have the disadvantage of limited height and position maneuverability compared to the pulley system. The pulley system is more maneuverable but is more difficult to adjust accurately.

10 Fans

Although a fan seems a simple machine, it has many parts which must function properly in order for it to work. The various parts are described in Figure 12. Use the figure as a guide to the following discussion about choosing a fan.

The specific task is to select a fan that will move the required amount of air against the resistance of the ventilation system. In choosing a fan, you should realize that every fan is designed to run most efficiently at a particular speed for a particular system resistance. Fan speed is measured in revolutions per minute, or RPMs. If you increase a system's overall resistance through changes in the work process, by letting material accumulate in hoods or in ducts or on the fan itself, or for any other reason, the fan's air capacity usually drops. Therefore you may have to speed up the fan, perhaps overstepping its motor's capabilities, or you simply might have to acquire a different fan.

Many factors determine choice of the right fan. Some are listed here, but keep in mind that what you want is a fan which uses the least amount of energy to move the most air, and one which can hold up under your projects. Most fans on the market have been given a certified rating performance seal of the Air Moving Contractors Association (AMCA).

Types of fans

The first and probably the most important decision to make about fans is to decide what type you need. Just as different types of hoods suit different situations, so various types of fans also have specific applications.

There are two general types of fans: the *axial fan,* where the airstream flows straight through the fan, and the *centrifugal fan,* where air is pulled into the center of the fan and is discharged perpendicularly to the fan's blades. (See Figure 13). Fans appropriate for artists' studios and shops are described below.

1. *Axial flow fan.* The most commonly used axial fan is the propeller fan. Propeller fans successfully provide cooling and dilution ventilation. They are very

good air movers because they move large volumes of air while using a minimum amount of electrical energy. They rarely are used in local exhaust ventilation systems because they can withstand only very little resistance.

Propeller fans, as used in a household window fan, a double window fan, or a large panel fan, can be used in a studio or shop to exhaust hot air or for dilution ventilation of solvent vapors. For example, if you operate a small electric enameling kiln in a room, a household window fan eighteen inches in diameter will give you some relief from heat and will dilute the metal fumes. As a rule, you can rely on a propeller fan to deliver 1,000 cfm air capacity for each square foot of fan surface area.

2. *Centrifugal fan.* Several different centrifugal fans are used in ventilation systems. The types differ in the way their blades are arranged on the fan wheel. The three most common wheels are the forward curved (FC), the backwardly inclined (BI), and the radial or straight blade.

The FC fan, also called a squirrel-cage fan, has many blades, all of them cupped in the direction of rotation. It is used for exhausting heat, for dilution ventilation, and for "clean" air (no particulates) local exhaust ventilation. *It should not be used for removing dusts, fumes, or condensible vapors* because those solid materials can accumulate on the squirrel cage blades, thereby reducing air flow.

Figure 12 Parts of a Centrifugal Fan

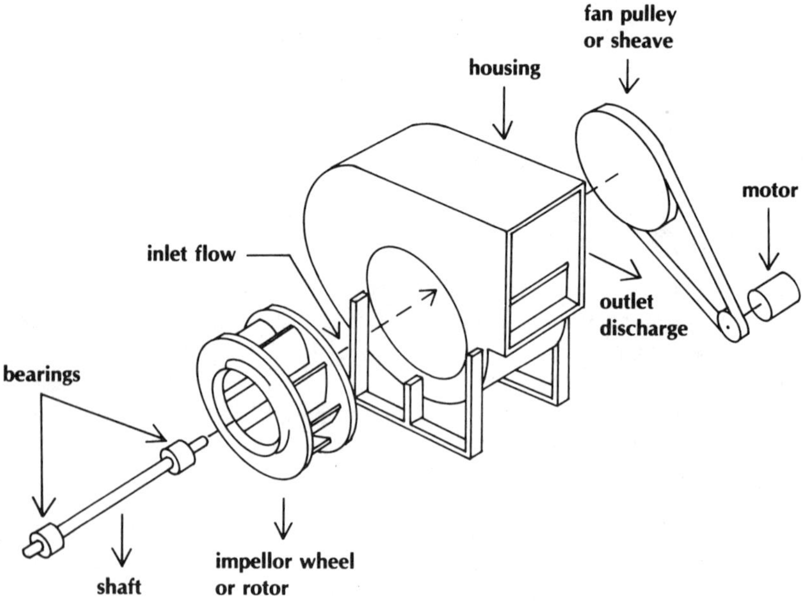

FC fans have low space requirements and low top speeds. And they are relatively quiet. Their mechanical efficiency is good. For a given air capacity (Q), their rotation speed is usually slower than other fans'. A drawback of the FC fan is motor burnout caused by electrical overloading. But because this fan is relatively inexpensive and widely available, one with a burned-out motor can be replaced easily. So an FC fan still may be a good choice when the alternative is a more expensive fan that does not burn out as quickly.

A BI fan has blades arranged in a configuration opposite the direction of rotation. It should be used mostly with clean air. However, it can be used for light dust loads, although its blades can become clogged with solid material, reducing air flow. BI fans have a higher number of revolutions per minute than FC fans. BI fans also tend to be "power limiting," meaning that they do not readily burn out their motors. One disadvantage is that they are difficult to find in small sizes (under twelve inches in diameter). Usually if you need a small fan you will find an FC fan satisfactory. BI fans can be noisier than FC fans. And for a given air capacity, they usually rotate up to forty percent faster.

Figure 13 Basic Fan Types

AXIAL FLOW

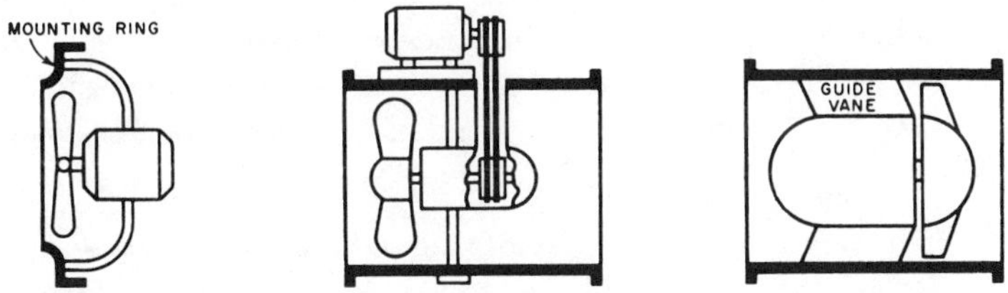

Propeller fan | **Tube-axial fan** | **Vane-axial fan**

CENTRIFUGAL

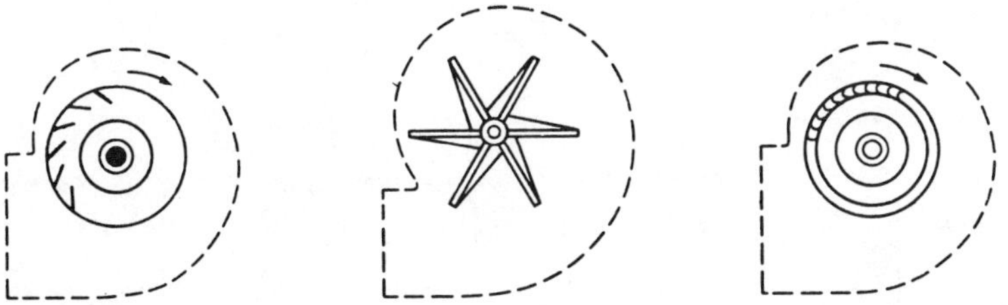

Backward curved blades | **Straight or radial blades** | **Forward curved blades**

Source: American National Standard Fundamentals Governing the Design and Operation of Local Exhaust Systems, ANSI Z9.2-1979 (New York: ANSI, 1979). Copies may be purchased from the American National Standards Institute at 1430 Broadway, New York City, 10018.

3. The blades of a *radial blade fan,* also called a *paddle-wheel fan,* are arranged perpendicularly to the fan shaft. The fan is used primarily in local exhaust ventilation systems and is often referred to as the workhorse of local exhaust ventilation. It is used to exhaust dusts and other solid particles. This fan can have a high rpm, be noisy, and be mechanically inefficient, but it is the only fan for dirty air.

Unfortunately selecting the right fan is not as straightforward as one might like. There are many situations in which any of several fans would be adequate (especially if you want to move clean air). Ventilation engineers spend many hours discussing fans and the gray areas in which one or another fan could be used. Some issues you should consider are listed here. If you still have questions, consult a qualified ventilation engineer.

1. *Volume of air* (cfm) required to be moved.

2. *Fan static pressure,* that is, the amount of resistence in the system which the fan can overcome.

3. *Fan efficiency.* Compare the amount of energy needed to run a fan with the volume of air moved. This will give you an idea of a fan's efficiency. The more energy needed to move a set volume of air, the less efficient the fan. Radial fans, for example, are very inefficient, whereas forward-curved fans are quite efficient. Some work processes require a specific type of fan, even though that fan might be less efficient than another. For example, a radial fan should be used in a woodworking shop to remove wood chips even though a radial fan is inefficient.

4. *Explosiveness or flammability of materials* that require non-sparking construction, including non-ferrous blades, and an explosive-proof motor and housing if the motor is placed in the air stream. (Refer to NFPA, NBFU, and local and state codes discussed in Appendix A.)

5. *Direct-driven versus belt-driven fans.* Speed can be adjusted on belt-driven fans, whereas there is only one speed for direct-driven fans. Direct-driven fans, however, take up less space and require less maintenance.

6. *Space limitations.*

7. *Noise.* Direct-driven propeller fans usually are quieter than the comparable belt-driven fan. And forward-curved fans are usually quieter than backwardly-inclined centrifugal fans.

8. *Operating temperature.* Fan bearings and other features may need to be able to withstand high temperatures. For example, sleeve bearings are suitable to

250° F, ball bearings to 550° F; special cooling devices are required at higher temperatures. Check with the manufacturer for specific information.

Table 10 compares the features of the fans described in this section. Additionally, a fan rating table is provided by individual fan manufacturers, repeating much of the information in Table 10—air capacity (Q), RPM (revolutions per minute), BHP (brake horsepower, the horsepower required to drive a fan), outlet velocity, and fan static pressure.

Table 10

Fan Selection

	Propeller fan (axial)	Forward curved fan (centrifugal)	Backwardly inclined fan (centrifugal)	Airfoil fan (modification of backwardly inclined fan)	Radial fan (centrifugal)
DILUTION VENTILATION	YES	YES			NO[2]
LOCAL EXHAUST VENTILATION	NO[1]	YES	YES	YES	YES
"CLEAN AIR"	YES	YES	YES	YES	YES
VAPORS, SOLVENTS, AND GASES	YES	YES	YES		YES[2]
DUST AND OTHER PARTICLES	NO	NO	FINE DUST AND FUME	NO	YES HEAVY DUSTS
SYSTEM RESISTANCE (STATIC PRESSURE)	LOW	LOW TO HIGH	LOW TO HIGH	LOW TO HIGH	LOW TO HIGH
ADVANTAGES	• Good for cooling and dilution ventilation. • Moves large volume of air against low resistance.	• Good for dilution and "clean air." • Inexpensive. • Widely available. • Low noise.	• Non-overloading characteristic makes this safest fan selection for varying resistance systems.	• Most efficient for very clean air.	• Most suitable fan for exhausting dust.
DIS-ADVANTAGES	• Not good for local exhaust ventilation. • Exhaust volume quickly decreases as resistance increases.	• Not to be used for dust. • Easily clogged by dust and sticky vapors. • Not power limiting.	• Not used for heavy dust.	• Should not be used for dirty air.	• Overall efficiency is lowest of all fan types.

1. Sometimes used in direct venting spray booths. 2. Not efficient but could work.

Fan motor

Select a fan motor that can provide the operating horsepower, including the brake horsepower (BHP) and a drive horsepower to run the fan. You will find brake horsepower values in fan rating tables. For small systems such as small studios and shops, a motor with a total horsepower at least twenty percent greater than the BHP should be used to power the fan. The motor you select must be able to run on the electrical service you have in your workplace.

Since your ventilation system may run for long periods, select a motor designed for continuous duty. Also, if access to your system's motor will be difficult, avoid motors that require frequent lubrication. If the motor will be exposed to dust or moisture, consult your fan supplier for advice. Motors can be obtained in drip-proof configurations, totally enclosed, or fully explosion-proof. You must inform your fan supplier about any special requirements, such as high heat or dust or moisture conditions.

Electrical service. Before you select a fan motor, check the electrical service where you work—the voltage provided (type of service) and the current, amperes, or amps available (size of service). Also check for any convenient electric lines and outlets. If possible, select a fan motor compatible with your existing electrical service. This stipulation may limit your choices. But if you choose a fan motor which will not operate on your electrical service, you may face extensive costly work by an electrician. Check with a licensed electrician and your power company for specific requirements. Work must be performed in accordance with all applicable codes. Inform your electrician if you plan to use flammable materials.

Most buildings and houses in the United States and Canada have sixty cycle alternating current. Direct current and fifty cycle (Hertz) alternating current are not usual north of the Rio Grande, but are common in other parts of the world. Utilities in the United States usually provide 240 volt, three-phase alternating current (except in New York City whose electric utility provides 208 volts). For residences, the utilities reduce this to single-phase, 120 volts at the meter, and you actually get about 110 to 115 volts at the motor. This is sufficient for many smaller fans and motors. Many buildings have three-phase, 240 volts (208 in New York City) at the meter, meaning you get about 220 volts (about 200 volts in New York) at the motor. More powerful motors needed by many industrial fans require this higher voltage. Because of these variations in electrical service, you must check with your local utility to learn what is available in your building.

Electrical wiring is limited by the amount of current (amps) it can handle without excessive overheating. For example the maximum current available on 115 volt AC (assuming there are no other loads on the line) with standard household wiring is about nine amps. The system should have fuse or circuit breakers to protect against excessive heating. Larger fans require heavier service and electrical wiring which can take the extra current. Check with a licensed electrician about proper wire sizes for your needs.

Connecting the motor. You sometimes can install small motors simply by plugging them in. A larger motor may have a built-in thermal overload device. These devices shut off power in the event the motor overheats. If the motor does not have a thermal overload feature, it may require a start/stop starter with overload protection for shutting off the power in case of an overload. If no outlet or line is convenient where you want the motor installed, you will have to arrange for one. This is a relatively minor and inexpensive electrical job. *Never connect a motor with an extension cord.* Extension cords are not meant to be permanent wiring. They can be damaged easily and cause fire and electrical shock hazards. For safety's sake, if you cannot see the motor from the place you start it, install a disconnect switch close to the motor. For example, if your fan is on a roof, a disconnect switch also should be on the roof. There are strict code requirements regarding disconnect switches, and for the safety of anyone working on your system, these codes should be followed.

Fan Installation

Where and how a fan and motor are installed can make a big difference in how well the fan performs. An improperly located fan even can result in contaminants being blown back into a workplace, rather than being removed. These guidelines may help you choose an optimum location for your fan.

1. *Fan location.* If possible, locate a fan downstream from the dust collector to minimize erosion and abrasion of fan parts.

2. *Fan inlet.* Eliminate elbows and other inlet obstructions if at all possible. Sharp elbows at the fan inlet seriously increase system resistance.

3. *Fan discharge.* Select a fan with a discharge outlet positioned so that exhausted air moves in the desired direction, thus eliminating additional bends and increased discharge resistance. (See Figure 14.)

4. *Fan arrangement.* There are ten numbered fan-motor-bearing arrangements for centrifugal fans. For use with corrosive, solvent, or dusty processes, the fan bearings must be outside the airstream. For these applications, arrangement one is the most common. For normal clean air applications, arrangement three is most common. Fans can be ordered with arrangements one or three (or arrangements for special applications). For your particular application, check with the fan manufacturer or supplier.

5. *Fan rotation.* Once you have decided on a discharge position, you must specify the direction of rotation from the drive side of the fan (see Figure 12) either clockwise or counterclockwise. (Figure 14 shows clockwise rotation.)

6. *Fan mounting.* Centrifugal fans on a roof should be mounted on an adjustable steel base attached to noncombustible supports. The base can come

Figure 14 Fan Discharges: Viewed from Drive Side

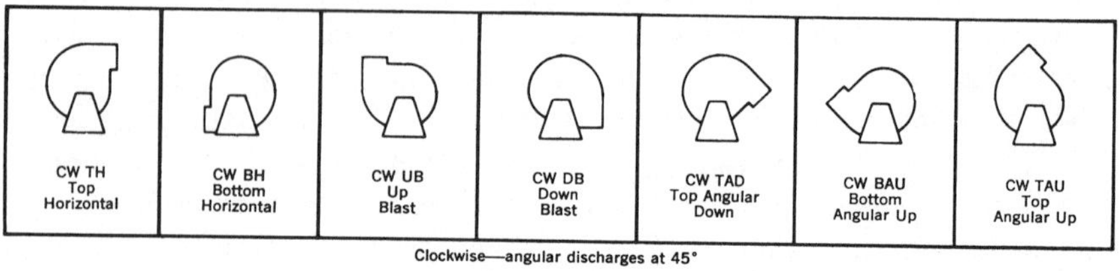

Clockwise—angular discharges at 45°

with or without vibration isolators. The fan manufacturer usually provides the mounting, if you request it. Leave at least eighteen inches of clearance between a fan discharge and the roof in order to keep the fan operating, even in deep snow.

7. *Noise reduction.* Mount fans on vibration isolators to reduce fan vibration and noise. Direct metal-to-metal vibration can be minimized by connecting a fan to a duct with a three-inch or wider flexible neck, which you can buy through sheet metal suppliers. As a temporary expedient, connect a fan and duct with two-inch duct tape. Be sure there is no metal-to-metal contact.

8. *V-belt fan drives.* Many fans are not direct drive, but are connected to a fan motor by means of a V-belt. The advantage of this arrangement is that fan speed can be altered by changing the ratio of the motor pulley to the fan pulley (often called the sheave). In many cases an adjustable pulley on the motor permits a range of adjustments to increase or decrease the motor pulley's effective diameter, which can change fan speed. Make adjustments on pulleys only when the motor is stopped and the electric disconnect switch is open. Make sure that the specified pulleys and V-belts are used for a particular fan, and that the V-belts are protected by belt guards.

11 Air cleaning devices

Air cleaning devices or collectors usually capture and remove materials that have made it through a ventilation system. The form of the materials being removed (fumes, gases, vapors, or dusts) will determine the type of collector needed. Although technical devices exist for collecting gases and vapors, they cost a lot and may be too costly for the average artist or small-business owner. Charcoal filters can collect many vapors and gases. Even though these charcoal filters are affordable they have the disadvantage of becoming saturated and ineffective unless they are maintained strictly and unless supplied with large amounts of activated charcoal. For these and other reasons, we do not recommend charcoal filters. Dust collectors, however, can be maintained easily, are affordable, and are adaptable to smaller operations. For these reasons, this manual concentrates on how an artist or small-business owner can choose the right dust collector.

Your ventilation system should have an air cleaner for three principal reasons: compliance with environmental regulations; removing particulates; and salvaging materials like precious metals that have a high resale value.

Choosing a dust collector

There are several things to consider when choosing a dust collector.

1. *Concentration and particle size of material.* Dusts are often described by particle size. Particle size is measured in microns (micrometers), designated by the Greek letter mu, μ. One micron (1μ) is equal to 1/25,000 of an inch. Particles range in size from the very fine (0.1 to 1μ), to large particles up to 100μ. Because people inhale particles less than 10μ in size into the deepest parts of their lungs, you should be most concerned about these smaller sizes. Figure 15 breaks down particle sizes according to specific processes which produce them. Dust concentration, particle size, and collection efficiency are major considerations when you decide what collector to purchase. Almost all collectors will do an adequate job for large particles. Some collectors can trap fine dust particles from operations

like fine sanding or buffing. However, few collectors remove extremely fine particles efficiently.

2. *Degree of collection required.* How much material a collector removes depends upon its "collection efficiency" for a specific type of dust. Collection efficiency is equivalent to the percentage of the material removed by the collector. The higher the collection efficiency, the greater the percentage of material taken from the airstream. Efficiency can vary with particle size, but efficiency and cost of the collector usually go hand in hand. How much material needs to be removed from the air depends upon several factors:

(a) Existing environmental standards.

(b) Nature of the contaminant. Does it have economic value, making it worth spending more money to collect a higher percentage?

(c) Is it a public nuisance? For example, will it corrode neighbors' property, or is there a strong odor associated with it?

(d) Is the cost of collection within reason?

3. *Characteristics of the airstream.* High temperatures as well as the amount of moisture in the air passing through a collector must be considered. Either one can ruin a filter, or at least decrease its collection ability considerably.

4. *Material properties.* Is there anything in the air which could damage a collector? Corrosive materials such as acids and alkalies, sticky materials such as metallic buffing dust from buffing compounds, and abrasive material such as sandblasting exhaust all can damage collectors. Collectors must be matched with the materials to be collected in order to prevent damage.

5. *Workplace space and collector size.* Dust collectors vary in size and shape. Be sure to keep the dimensions of your workspace in mind when you choose a dust-collection unit.

6. *Methods of disposal.* How will collected material be disposed? For dusty operations, the cleaning and disposal of the collected dust often create a secondary problem.

7. *Noise.* The location of some types of dust collectors can be influenced by noise factors.

Types of dust collectors

There are two types of dust collectors that are both economical and suitable for art processes and other small operations: cyclone collectors and fabric collectors. Both are used widely in industry and are available in various sizes from commercial suppliers. Another type of dust collector, electrostatic precipitators, has more limited use.

Cyclones, also called centrifugal collectors, remove coarse to medium-sized particles by spinning the air in which they are contained in a cone-shaped chamber. The particles are thrown to the edge of the spinning airstream, then settle out. This method will not effectively remove very fine particles. Cyclones are particularly useful in woodworking shops, where most dust particles fall in the large-to-medium range. Most large-body cyclones have separate fans. You can place the bodies inside a shop, with their fans outside or on the roof so that they do not create a noise problem. Some medium-body cyclones have integral fans, not separate ones. These are very efficient but if they are installed inside a shop they may produce excessive noise. Cyclones require low maintenance and they are low in cost. They are also very durable and can withstand heavy use.

Fabric collectors, also called "bag houses," are highly efficient filters for medium-sized to very fine particles. They remove dusts by passing air through layers of fabric at low air velocities. Gradually a dust layer is produced on the fabric, and this layer acts as the filter. The fabric merely supports this dust layer so that in effect the collected dust itself becomes the dust-collecting agent. After prolonged or heavy use, the fabric becomes clogged and resistance through the collector increases unduly. Then the fabric must be cleaned or replaced. Fabric collectors can be damaged easily, and tears and rips allow collected dust to escape. So if you conduct other operations that could cause a rip, or if vandalism is a problem (in a school, for example), you may find a fabric collector an unwise choice. Fabric collectors require regular cleaning and maintenance.

To improve a ventilation system's overall collection efficiency, two types of dust collectors sometimes will be used in series. For example, a cyclone filter can be followed by a fabric collector. You would find it cost effective only if you do a high volume of work.

Occasionally you can use a vacuum cleaner as a portable dust collector. If one of your operations generates a very small amount of nontoxic and nonflammable dust and if you undertake this process infrequently, you may find a common vacuum cleaner adequate for the job. You must position it close enough to capture the dust particles. An appropriate vacuum cleaner must be selected to handle the specific particles in order for this method to be effective.

If you use a vacuum in a woodworking shop or any other area where flammable and combustible dusts are produced, make sure that it is equipped with an explosion-proof motor to guard against fire and explosion.

A final caution about using vacuum cleaners. Never use them with gases and vapors. Vacuums cannot remove gases or vapors unless they are equipped with a special collection device. If you draw gaseous materials through a vacuum cleaner, you could recirculate toxic air. You might damage the vacuum's motor or other parts, and you might cause a fire or explosion.

The third type of dust collector, electrostatic precipitators (ESP), removes fine particles and fumes from air by placing an electrical charge on the particles, then attracting these particles to an oppositely charged collection plate. Electrostatic precipitators usually are mounted on the ceiling or placed on a table, drawing in contaminated air and recirculating clean air. They are highly efficient for fine

Figure 15 Range of Particle Sizes of Common Dusts and Air Cleaner Collector Performance

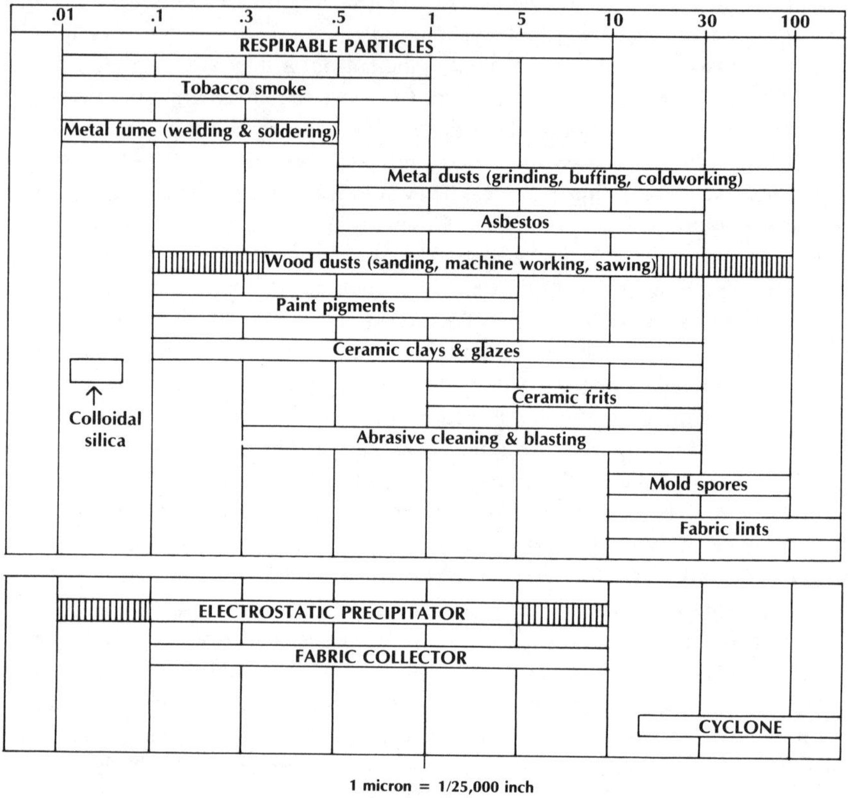

Source: Stanford Research Institute, *ACGIH Ventilation Manual.*

particles but cannot be used for dusty operations because they clog easily. Do not use ESPs for highly toxic materials because ESPs can recirculate them into a room's atmosphere. Nor should they be used around flammables or in a room that requires explosion-proof conditions. They can touch off explosions because electrical arcing cannot be prevented. Even though electrostatic precipitators can effectively remove fumes or fine particles from air, they usually cannot remove vapors or gases. The recirculated air may look clean but these devices actually return toxic gases and vapors continually to the workplace. In photographic darkrooms, for example, ESP units remove dust or smoke from the atmosphere and thereby improve the photographic process itself. But they do not remove the highly irritating sulphur dioxide gas that is generated during developing. ESP's are generally expensive, and they require frequent maintenance.

There are many cheap, so-called "air purifiers" on the market which make extravagant claims that they can remove toxic substances and odors from the air. Most of these are related to electrostatic precipitators in that they place a charge on particles. However, they usually do not have a collector and simply emit the charged particles into the air where they coat nearby surfaces. Even the ones with collectors are not suitable for studios because their collecting capacity is very small. These air purifiers with collectors might be adequate for cigarette smoke, house dust, and pollen but should not be used with toxic dusts.

12 Final precautions

While the system is being planned and after it has gone into operation, there are a number of things that can be done to make the system safer and more effective.

Protecting against fire and explosion

Flammable and combustible materials, if allowed to form an ignitable mixture in the air and in the presence of an ignition source, can cause an explosion or fire. If you work with materials such as flammable solvents, combustible organic or metallic dusts, or easily ignitable fibers, you should consult local fire prevention codes. The codes will tell you how to install ventilation and electrical equipment safely. Most of these codes are based on the National Fire Protection Association (NFPA) standards and the National Electric Code (NEC). (More complete references appear in Appendix A.)

These codes probably will apply to your operation if you use a ducted ventilation system (a spray booth, for example) to remove flammable vapors, or if you use and store large amounts of flammable liquids that might spill, or if you work with any flammables in enclosed spaces. If you are unsure whether your project produces an explosive atmosphere or a potentially explosive one, you may want to consult either a reputable ventilation engineer or an industrial hygienist. Either can give you an initial evaluation and help you decode the appropriate codes. If you learn that the codes do indeed apply to your work, you will need these experts to design and install a ventilation system that complies with the law.

The NFPA and NEC requirements are based on the principle of keeping ignitable mixtures completely isolated from heat and spark sources. Two features are needed to control explosive atmospheres:

1. *Explosion-proof construction of ventilation parts.* Seal all joints and seams in the hood and ductwork to prevent vapors from escaping into workspace air. Use spark-resistant fan parts made of non-ferrous metal, such as aluminum or bronze.

2. *Install explosion-proof electrical equipment.* This will prevent contact of flammable materials with sparks or heat-producing electrical parts, including light fixtures, switches, wiring, circuit breakers, and motors. Tools used to clean, repair, or maintain the ventilation system and other systems should be made of non-sparking metal to prevent their igniting flammable materials.

Checking the system

After a ventilation system has been installed, you should have it checked to see that it is operating properly. If an engineer, industrial hygienist, or installer worked on the job, he or she should make an initial check and recommend any changes necessary to meet design specifications. If you did not consult with an expert, you may decide to consult one at this stage. Tests to evaluate ventilation systems require special equipment and skills. For example, air flow devices are used to measure air velocity and determine air capacity in a system's different parts. Air sampling instruments are used to measure the level of contaminants in the air. So final checks obviously are best made by persons with the right knowledge and equipment.

1. *Keep records.* Keep a permanent record about the specific test methods used and their results. Initial test results then can be used as a baseline for future comparisons. You must check the system again whenever additional hoods are contemplated or when you change a process. Ventilation measurements also may be necessary when a system does not work properly and you need to locate where the problems are.

2. *Common-sense observations.* Without special equipment, you can make common-sense observations about the performance of your system. Remember, the purpose of ventilation is to remove contaminated air from the workspace so the best test of a system is whether the air is safe to breathe. While your process is underway and you have the ventilation system on, step back and use your senses.

(a) Can you see dusts or mists in the air or settling on nearby surfaces?

(b) Can you smell any gases or vapors?

(c) Do you feel any eye, nose, or throat irritation, or have you any other symptoms?

If your senses give you positive answers to any of these questions, then contaminated air is escaping the draw of the hood and the air may not be safe to breathe. Unfortunately, many dusts, fumes, gases, and vapors are odorless, non-irritating, and invisible to the naked eye. Therefore, these simple observations are useful only to prove a system unsafe. Not seeing or smelling chemical contaminants does not mean the system is working properly.

3. *Checking the hood.* You also can check the air flow into the hood even if you have no special equipment or skills. Smoke tubes that emit a highly visible smoke are made for this purpose. When smoke is released across the face of the hood and at the source of the contaminants, the hood should draw the smoke completely and quickly from the area. If the smoke is not captured completely or if it dissipates, then the hood is not working properly.

You can acquire smoke tubes from safety equipment suppliers for about $50. If you anticipate frequent tests or if you have to check many hoods, you will find smoke tubes are a worthwhile investment. However if you will not be checking your hood often, a lit cigarette or incense will produce enough visible smoke to enable you to make a good observation. Be sure you do not use this method if there are any flammables in the area. Also, children's soap bubbles can be blown across the hood to check for effective capture.

4. *Troubleshooting.* If any of these tests indicate that your system is not functioning properly, here are a few things to consider.

(a) Is the fan belt worn or slipping? Is the fan wired correctly into the electrical service? Is the fan wheel rotating in the correct position? (Backwards installation of centrifugal fans will cause a decrease in air capacity.)

(b) Is the air cleaner clogged or in need of replacement?

(c) Is there an obstruction in the ductwork caused by a closed damper or accumulation of material?

(d) Has adequate make-up air been supplied?

If you cannot locate the problem yourself, you will have to consult an expert. The sooner you do this, the sooner your system will operate properly and do the job you want it to do.

Where to go for help

Depending on the type of operation you have, the space and money you have to work with, and your own abilities, you may need help from one or more experts. This manual gives you much of the information you need to select and install a ventilation system suited to your needs. The manual also will guide any experts whom you consult. Five different types of specialists are involved in ventilation:

1. *Registered professional engineer.* If you need an industrial ventilation engineer's services, you must find one who is capable of evaluating your problem and designing a good system. Consult your local engineering societies, Yellow Pages, or other directories.

2. *Heating, ventilating, and air conditioning (HVAC) engineer.* HVAC engineering outfits are more numerous and available than industrial ventilation engineers. If their staffs consult this booklet, many good HVAC businesses ought

to be able to design and install a system for you. Check directories and Yellow Pages, and also ask friends and associates for referrals.

3. *Industrial hygienist.* You may decide that your workspace environment needs to be evaluated for toxic dusts and gases. In this case, you must consult an industrial hygienist to sample the air. Check with your state health or labor departments for consultants. States and some municipalities provide this service free. Also check with any local committees on occupational safety and health or with the Center for Occupational Hazards for names of consultants.

4. *Sheet metal works.* Ventilation systems are made from sheet metal. You probably will need the services of a sheet metal shop to build all or part of your system. Consult Yellow Pages under sheet metal, blow pipe, or ventilation.

5. *Electrician.* Ventilation system installation may require some electrical work and adherence to local fire and building codes. Compliance with codes requires that a licensed electrician do the work. The best way to find a licensed electrician is by asking friends and associates for referrals. Otherwise check the Yellow Pages. Choose an expert who best suits your special needs.

III

Five specific ventilation-system designs

This section describes ventilation systems designed for several processes and situations commonly used by artists: a ceramic kiln, spray painting booth, and acid-etching, welding, and woodworking shops. A step-by-step design procedure is presented for each. Familiarize yourself with the general information in the first two sections of this handbook before you study the individualized design specifications in this chapter.

Use these specifications only as a guide for designing and developing your own system. Each design was made for a specific operation that requires a particular hood type, duct run, duct size, and fan. Duplication of any design may or may not provide adequate ventilation for your process because certain assumptions have been made which may not apply to you and your work. Because of the differences between processes and working conditions, each situation requires certain adaptations and changes. All the systems discussed here are assumed to comply with local, state, and federal codes and regulations. Be sure to study the actual regulations when you design your own system.

13 Canopy hood design: electric ceramic kiln

During a kiln firing, a ceramicist can be exposed to a wide range of toxic gases and fumes from clays and glazes, as well as to deadly carbon monoxide gas if the kiln is gas-fired. A properly designed canopy hood can minimize these exposures.

The following specifications have been designed for a three-foot-diameter, top-loading electric kiln. The hood will be cone-shaped and four feet in diameter. The hood will be positioned seven feet from the floor to allow sufficient headroom for work. (See Figure 16.)

For a small kiln used moderately, an air capacity (Q) of 100 cfm per square foot of hood face is assumed to be adequate in most cases. The surface area of the four-foot-diameter hood face is 12.56 square feet. (See Table 12 of Appendix B for areas of circles.)

$$Q = 12.56 \text{ ft.}^2 \times 100 \frac{\text{cfm}}{\text{ft.}^2} = 1{,}256 \text{ cfm.}$$

Normally, this amount is rounded off to 1,250 cfm.

Ducts

For this operation, an economical duct velocity of 2,000 fpm is reasonable. (Refer to Table 9 for recommended duct velocities.) To determine the duct size, use the air capacity formula

$$Q = AV:$$
$$1{,}250 \text{ cfm} = A \times 2{,}000 \text{ fpm}$$
$$A = 0.625 \text{ ft.}^2$$

Referring to Appendix B, you will find that 0.625 ft.^{2} falls between a ten-inch duct (0.545 ft.^{2}) and a twelve-inch duct (0.785 ft.^{2}). Both duct sizes are available commercially, but the twelve-inch duct is preferable because it is larger and therefore has lower resistance and is more economical to use.

Because this decision changes the duct area, the duct velocity must be recalculated. Using Q = AV, you find:

1,250 cfm = 0.785 ft.^{2} × V

V = 1,592 fpm

This is rounded off to 1,590 fpm.

Assume this duct will run from the top of the canopy hood to a fan, then to a panel in the window of an outside wall. Discharge will be through a back draft damper (sometimes called a gravity shutter). Because of the physical environment, two ninety-degree bends are necessary. (See Figure 16.) The total length of straight duct is fifteen feet.

System resistance

The overall resistance of the system is found by adding the separate resistances of the system components. How to compute the resistance in this system is illustrated in Appendix B. The final computations are:

Components	*Resistance (inches of water)*
Canopy hood	0.18
Ducting	0.20
Air cleaner—none	0.0
Back draft damper (assumed)	0.10
Total	0.48

Fan selection

The fan static pressure must be calculated in order to select a fan capable of operating the system. Fan static pressure is defined on page 52. The calculations for this system are illustrated in Appendix B. The fan static pressure is calculated to be 0.31 inches of water, which is equal to approximately 3/8 of an inch of water. (Fan static pressure is generally expressed in eighth-inch increments.)

The process for selecting the proper fan for this system and the necessary calculations are also illustrated in Appendix B. The final fan data for selection and purchase of the fan are:

Type: backwardly inclined centrifugal fan
Size: 12$\frac{1}{4}$ inches, also called a 122
Air capacity: 1,250 cfm

Static pressure: 3/8 inches of water
Fan speed: 1,432 revolutions per minute (RPM)
Brake horsepower (BHP): 0.22

Either a 1/3 or 1/2 horsepower, single-phase motor could be selected to run the fan. The large motor is a safer choice and allows for design inaccuracies.

Figure 16 Canopy Hood Exhaust System

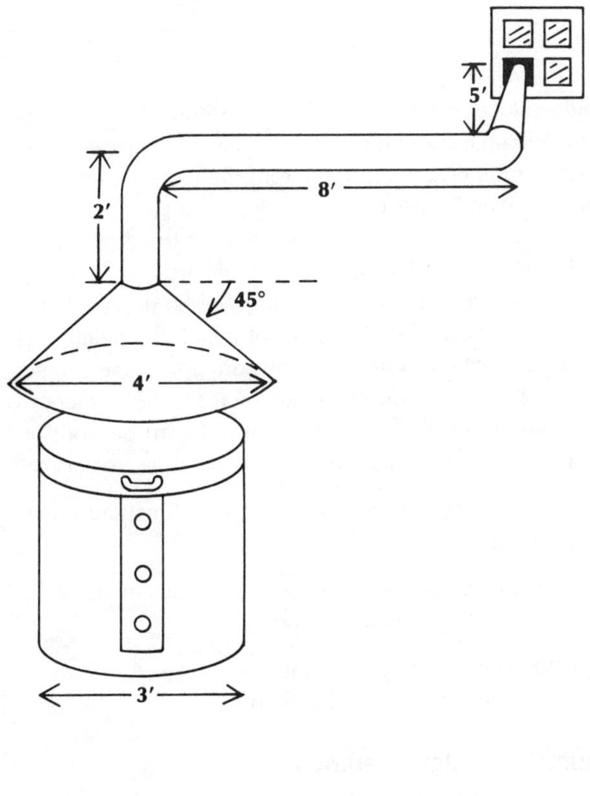

14 Slot hood design: acid etching

Acid etching produces acid fumes which can burn the nose and lungs, causing acute or chronic damage to the respiratory system. A properly designed slot hood will remove these fumes, taking them from the worker's breathing zone and out of the workplace. (See Figure 17.)

In this example, the acid-etching process is carried out on a table forty-eight inches long by thirty inches wide, located against a wall. A slot hood forty-six inches long with a forty-three-inch opening will be designed. This measurement makes good use of a standard size of sheet metal (forty-eight inches) and leaves allowances for connection to the hood plenum and at the slot edge. In addition, side baffles, or side shields, would greatly increase the efficiency of this design.

For a slot hood, an air capacity (Q) of 350 cfm per foot of hood length is acceptable. (Do not use this figure if the table depth is greater than thirty inches.)

$$Q = 350 \text{ cfm/foot of hood length} \times 4 \text{ feet (approx.)}$$
$$= 1{,}400 \text{ cfm}$$

A suggested slot velocity for acid gases and solvent vapors is 1,000 fpm. To determine the total slot opening area, use $Q = AV$:

$$1{,}400 \text{ cfm} = A \times 1{,}000 \text{ fpm}$$
$$A = 1.4 \text{ ft.}^2 = 201.6 \text{ in.}^2$$

Size and number of slot openings

Knowing the slot length, forty-three inches, and the area of the opening, 201.6 square inches, the slot height can be calculated from $A = L \times H$

$$201.6 \text{ in.}^2 = 43 \text{ in.} \times H$$
$$H = 4.7 \text{ in.}$$

The total area of the opening is 201.6 square inches (43 × 4.7 inches). To improve the hood's capture efficiency, this total area is divided into five slots, each 15/16 inches by forty-three inches. (The number of slots can vary according to the height at which work will take place.) The slots are spaced with 1½-inch clearance at sides, top, and bottom of the hood, as shown in Figure 17. The hood is positioned on the wall eight inches above the surface of the table to accommodate the acid trays. This height can be varied to suit the situation. To accommodate the increase in air capacity from the first to the fifth slot, the plenum face could be slanted. Or you may use dimensions that will allow economical fabrication from sheet metal.

Figure 17 Slot Exhaust System

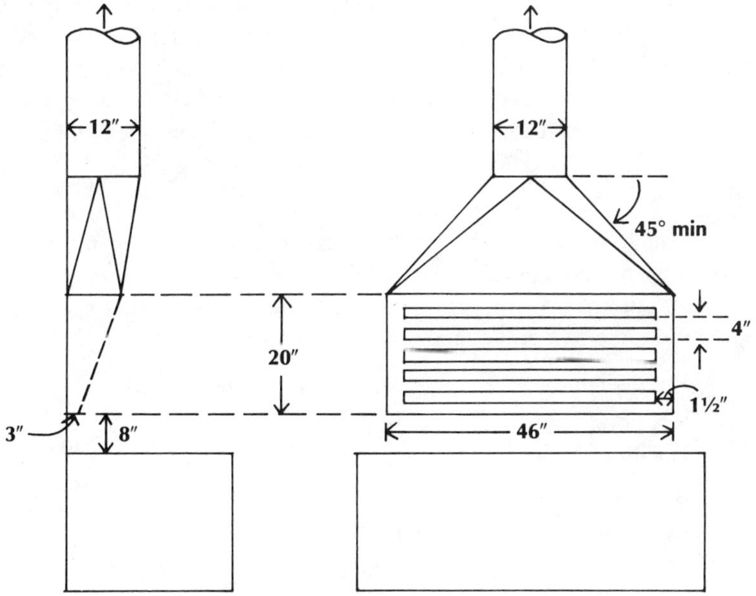

Ducts

A suitable duct velocity for this system is 2,000 fpm. (Refer to Table 8 for recommended duct velocities.) To determine the required duct size, use the air capacity formula $Q = A \times V$:

$$1{,}400 \text{ cfm} = A \times 2{,}000 \text{ fpm}$$
$$A = 0.70 \text{ ft.}^2$$

Referring to the chart in Appendix B, you will note that 0.70 ft.^{2} falls between an eleven-inch duct (0.66 ft.^{2}) and a twelve-inch duct (0.7854 ft.^{2}). The twelve-inch duct is chosen because it is larger and therefore has a lower resistance and is more economical to use. Because the duct area has been changed, the duct velocity must be recalculated. Again using $Q = AV$, you will find:

$$1{,}400 \text{ cfm} = 0.79 \text{ ft.}^2 \times V$$
$$V = 1{,}783 \text{ fpm}$$

This is rounded off to 1,785 fpm.

This studio has a flat roof, so the ductwork can travel straight from the hood up and out the roof. At the roof, the duct will take a ninety-degree bend and run horizontally to the exhaust fan. The vertical length of the duct from the hood to the bend is five feet, and the horizontal length of the run along the roof is also five feet. Total length of straight duct is ten feet with one ninety-degree bend.

System resistance

The overall resistance of the system is computed by adding the separate resistances of the system components. Appendix B includes an illustration which computes the resistance for the canopy hood design. Refer to that example for the method used to arrive at the final computations shown here.

Component	*Resistance (inches of water)*
Slot hood	0.36
Ducting	0.13
Air cleaner (not used)	0.00
Discharge (assumed)	0.10
Total	0.59

Fan selection

The fan static pressure is 3/8 of an inch. Refer to Appendix B for an example of computing fan static pressure.

Knowing the fan static pressure, you now can refer to the manufacturer's fan chart to choose the correct fan. Either a forward curved fan or a backwardly inclined fan would be appropriate for this situation. Since the fan will be small and on an exposed roof, a commercial utility fan, which has a motor and drive in one assembly and is weather covered, will be adequate. Using a manufacturer's fan chart for a forwardly-curved or backwardly-inclined fan, you will find that either fan sizes 122, 135, or 150 could deliver 1,400 cfm with a 3/8-inch fan static pressure. While size 122 could do the job, do not choose it because it has the highest brake horsepower. Either the 135 and 150 sizes would be adequate. The smaller 135 has the advantage of being slightly less expensive and matches the design specifications for this system more closely.

Once you have determined the size of the fan, it is a good idea to consult the manufacturer or his or her representative, asking exactly which fan will serve your special situation. In this case, the 135 fan has a seventeen-inch fan inlet, requiring an enlargement of the twelve-inch duct near the fan to seventeen inches. Recalculating the fan static pressure and Q, using a seventeen-inch diameter size, you will find that the new fan static pressure is 0.51 inches of water, which is closer to a fan static pressure of half an inch of water. Referring to the manufacturer's tables, the final fan data for selection and purchase are:

Type: forward curved utility fan
Size: 13½ inches, called a 135
Air capacity: 1,462 cfm
Static pressure: ½ inch of water
Fan speed: 1,237 revolutions per minute (RPM)
Brake horsepower (BHP): 0.19
Motor: 1/3 HP

The motor's phase (single or three) and your available voltage will have to be specified. In this instance it is likely you will want a 120-volt, single-phase, 1/3 horsepower motor.

15 Spray booth: spray painting

Spray painting and other types of spraying produce large amounts of potentially toxic vapors and particulates. If the spray is solvent-based, then a fire and explosion hazard exists, too. A properly designed spray booth ventilation system can minimize any hazard. (See Figure 18.)

This spraying operation is performed in a third-floor studio in a six-story building. Because of the size of the objects to be sprayed, a four-foot-by-four-foot, bench-type spray booth was purchased. The working depth of the booth is three feet. The booth's air cleaner consists of metal baffles covering ninety percent of the target area. To conform with local fire and building codes and to avoid contaminating others' lofts or apartments and the like, the spray booth has a ductwork exhaust running out the rear of the building and up to the roof, and it has a rear discharge outlet.

The usual air capacity requirement is 125 cfm per square foot of booth opening. So, $Q = 4 \text{ feet} \times 4 \text{ feet} \times 125 \text{ cfm} = 2{,}000 \text{ cfm}$ minimum.

Placement of baffles in the rear of the booth breaks the booth into essentially two hoods. In the example, the baffles occupy ninety percent of the surface area, leaving ten percent of the area open for air to pass through. In this case the velocity through the slot (baffles) will be higher than in the ducts. To determine the velocity through the baffle slot, use $Q = AV$:

$$2{,}000 \text{ cfm} = 10\% \, (4 \text{ feet} \times 4 \text{ feet})\, V$$
$$V = 1{,}250 \text{ fpm}$$

Ducts

The spray booth is equipped with an eighteen-inch duct rear outlet. Use $Q = AV$ to find the duct velocity:

$$2{,}000 \text{ cfm} = 1.767 \text{ ft.}^2 \times V$$
$$V = 1{,}130 \text{ fpm}$$

Figure 18 Spray Booth Exhaust System

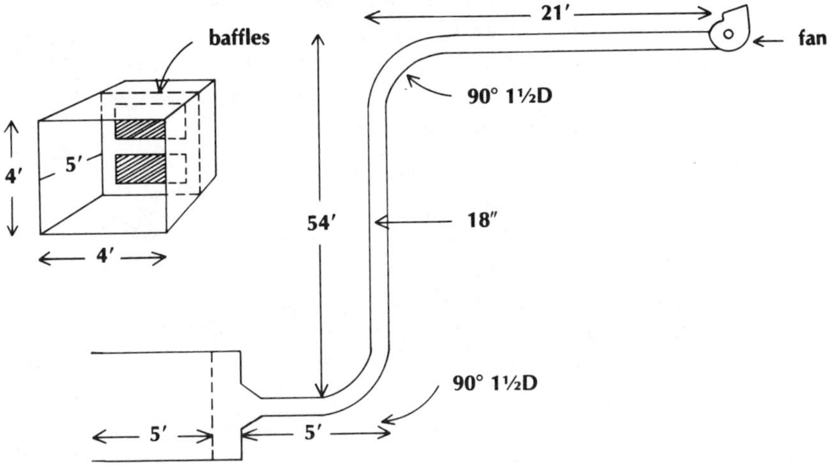

The spray booth is located at the rear of the studio, close to the outlet for the ductwork. The fan is placed on the roof. The duct runs five feet to the rear window replacement panel, makes a ninety-degree, one-and-a-half-inch diameter bend and runs fifty-four feet straight up; makes another ninety-degree bend to the roof and finally a twenty-one-foot straight run to the fan. The fan discharges upward through a no-loss stackhead. Total length of straight duct is seventy-nine feet.

System resistance

The overall resistance is computed by adding the separate resistances of the system components. See Appendix B for an illustration computing the resistance for the canopy hood design. Refer to that example for the method used to arrive at the final computations shown here for the spray booth design.

Component	*Resistance (inches of water)*
Spray booth	0.139
Air cleaner (baffles)	0.178
Ductwork	0.161
Discharge, no loss	0.0
Total	0.478

Fan selection

The fan static pressure is half an inch. Refer to Appendix B for an example about computing fan static pressure.

Because the discharge air is relatively clean, a forward curved Junior Utility Fan with a V-belt drive is selected. It is equipped with a weather cover, since it is exposed on the roof. There is a start/stop switch at the booth and an electrical disconnect on the roof. All electrical work was done to conform with local codes. The final fan data taken from the manufacturer's fan table are:

Type: Junior Utility Fan (non-sparking construction)
Size: 15½ inch, called a 155
Air capacity: 2,047 cfm
Static pressure: ½ inch of water
Fan speed: 554 revolutions per minute (RPM)
Brake horsepower (BHP): 0.44
Motor: 3/4 horsepower (HP)

16 Local exhaust system: woodworking

Woodworking shops in schools or individual studios are complicated since they usually have several woodworking machines, each requiring local exhaust ventilation to remove wood dust. Each machine must have its own hood and duct, leading first to a main duct, then leading in turn to a dust collector and fan. This type of system usually is designed by an engineer. The example given here illustrates the complexities of this kind of exhaust system and will help you understand what information you need when you consult a ventilation engineer. The example is based upon an actual school situation.

The shop contains the following woodworking machines: (1) a small jointer, (2) a disc sander with fifteen-inch disc and attached collector hood with a $2\frac{1}{2}$-inch outlet, (3) table saw with a ten-inch circular blade, (4) band saw with a fourteen-inch wheel and 5/8-inch blade, (5) spindle sander with cylinder four inches by eight inches, (6) drill press, and (7) tool grinder. The principal hazards they create are fine wood chips from the table saw and fine sawdust from the disc sander. The drill press and tool grinder are used infrequently, so they are not being considered. Housekeeping is the solution to any dust problem they create.

Collector hoods

The first design step is to decide the type of hood, exhaust volume, and duct velocity for each machine. The required information to make the decisions can come from prior experience, manufacturers' recommendations, or from design specifications in *Industrial Ventilation: A Manual of Recommended Practice*. Several of the most-used design specifications from this manual are reprinted at the end of this section (figures 20-25). For most shops having small woodworking machines, use the lowest exhaust volumes specified. If your machine is larger, use the appropriate volumes indicated. If there is a contradiction between the manufacturer's recommendations and those in the industrial ventilation manual, choose the one with the highest exhaust volume.

Table 11 lists the specifications for exhaust volumes, slot and duct velocities, and calculated duct diameters for the various machines in this example. The existing hood for the disc sander will have to be discarded because it has an outlet $2\frac{1}{2}$ inches in diameter, whereas the duct diameter calculated from the exhaust volume and duct velocity comes to $4\frac{1}{2}$ inches. (Using the $2\frac{1}{2}$-inch outlet would give a duct velocity of 12,941 fpm.)

Small band saws often have only one hood at their bases. But the one in Figure 23 has two hoods. The value chosen is for the sum of the two hoods.

There is no design specification in *Industrial Ventilation: A Manual of Recommended Practice* for the spindle sander, and the value chosen derives from experience. Also consult its manufacturer.

If all the machines will be operating at one time, then a total exhaust volume of 2,190 cfm is required. If only one or two machines are used at a time, then the full 2,190 cfm would not be needed, and blast gates (like dampers) could be used to close off ducts of unused machines. The problem with blast gates is that they are very susceptible to changes in position, which could disrupt the air flow in the system.

Since this example concerns a school shop, it can be assumed that all machines might operate at the same time. Therefore an exhaust volume of 2,190 cfm should be used.

There are three possible collectors for woodworking shops: (1) a large body cyclone, (2) a small body cyclone with attached fan, and (3) a fabric collector. Their advantages and disadvantages have been discussed previously but here is a summary. The large-body cyclone is inexpensive, has low resistance losses, but it is not efficient with fine sawdust and it is bulky. The small-body cyclone is more efficient, but its attached fan can be very noisy. The fabric collector traps fine dust most efficiently and is quiet, but it is more expensive, requires more maintenance, is subject to rips and possible vandalism, and its fabric filter has to be replaced every few years. Because this illustration is based upon an actual school woodworking shop, and since that actual shop has not been adequately maintained by school staff, it would be wise not to select a collector that requires regular maintenance or one subject to vandalism. Taking all of these things into consideration, it was decided to use a large-body cyclone with a fan on the roof. The cyclone is installed in the shop for ease of maintenance; the fan is put on the roof to reduce noise irritation. Recirculation of the air was ruled out because some fine dusts might re-enter the room.

Table 11

Wood Dust Collector Hood Specifications

Woodworking machine	Size	Exhaust volume	Slot velocity	Duct velocity	Duct size
1. Jointer (Rockwell Delta)	small (less than 6″)	350 cfm	2000 fpm	3500 fpm	4″
2. Disc sander (State Mfg. Co.)	15″	440 cfm	2000 fpm	3500 fpm	4½″
3. Table saw (Rockwell Delta)	10″	350 cfm	2000 fpm	3500 cfm	4″
4. Band saw (Rockwell Delta)	⅝″	700 cfm	—	3500 cfm	6″
5. Spindle sander (State Mfg. Co.)	4″ × 8″	assume 350 cfm	2000 fpm	3500 cfm	4″

Ducts and system resistance

Plan the placement of machines and the cyclone, if at all possible, where you can minimize the amount of ducting. Figure 19 shows the location of two of the machines and the cyclone. Make sure the ducts do not interfere with operation of the machines. When possible, use short lengths of ducts to connect machines to a major duct. The main duct then goes to the cyclone, then from the cyclone to the roof where the fan and discharge are located.

Duct sizes are calculated from the exhaust volume and duct velocity for each machine. The calculations for duct resistance start at any machine and duct branch, then progress to the cyclone. They are calculated as in Appendix B. Determine duct and hood resistances for each hood and duct branch up to the point where two branches meet. Note that the pressures at the junction of two ducts must be equal. If they are not, then the air flows in each duct will vary from the calculated values. This is called a balanced duct system and requires special calculations. Adjustments and recalculation of duct sizes and velocities must be made if the static pressures to branch junctions differ by more than twenty percent. If you make these calculations for all the hoods and duct branches, cyclone, duct from cyclone to fan, and the discharge you will obtain the following:

Component	*Resistance (inches of water)*
hoods and ducts leading to cyclone	4.26
large-body cyclone (assumed)*	0.50
duct from cyclone to fan	0.36
discharge resistance (assumed)	0.10
Total	5.22

Fan selection

The fan static pressure is five inches of water (see Appendix B for a sample calculation) for a fifteen-inch duct with an exhaust volume of 2,190 cfm and 1,770 fpm duct velocity. A duct velocity of 3,500 cfm is not needed for the duct leading from cyclone to the fan since the air is relatively clean and does not have to transport dust.

*This resistance can be as high as two inches for certain large body cyclones.

Figure 19 Woodworking Shop Dust Collecting System

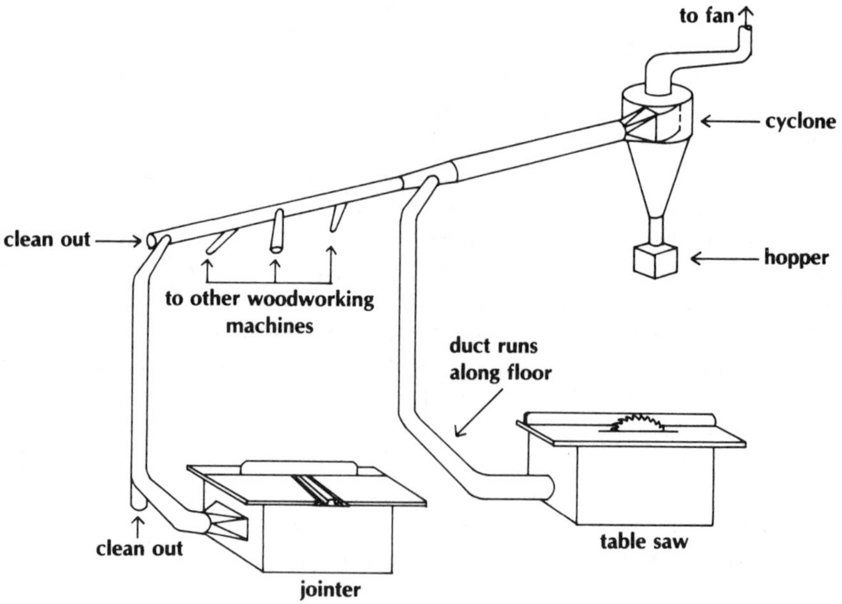

A forward curved or backwardly inclined centrifugal fan could be used for this situation. The fan table of one manufacturer lists a size 122, class two backwardly inclined centrifugal fan with wheel diameter of $12\frac{1}{4}$ inches which has an exhaust volume of 2,184 cfm, fan speed of 3,413 rpm, and brake horsepower of 2.88.

For a fan of 2.88 BHP, a five HP motor could be selected. Since the school has adequate electrical service, the motor could be three-phase, 220 volts (or 200 volts in some cases). Since it is to be located on the roof, an electrical disconnect switch also should be placed on the roof nearby.

The final fan data taken from the manufacturer's fan table are:

Type: backwardly inclined centrifugal fan
Size: $12\frac{1}{4}$ inch, called a 122
Air capacity: 2,184 cfm
Static pressure: 5 inches
Fan speed: 3,413 RPM
Brake horsepower (BHP): 2.88
Motor: 5 horsepower (HP), 3 phase, 220 volts

For an exhaust volume of 2,190 cfm, make-up air would likely be required to replace the exhausted air. This could be provided by an overhead make-up unit with steam heating. The amount of make-up air would not have to equal exhausted air since there is some natural leakage. A make-up unit providing 2,000 cfm should suffice. The make-up unit should have standard throw-away filters before the steam coils in order to minimize maintenance and clogging of steam coils. Possible alternatives to the steam-heating coils could be gas-fired heaters, electric coils or hot water coils.

Figures 20 to 25 appear on the next three pages.

Figure 20 Jointer Dust Collecting System

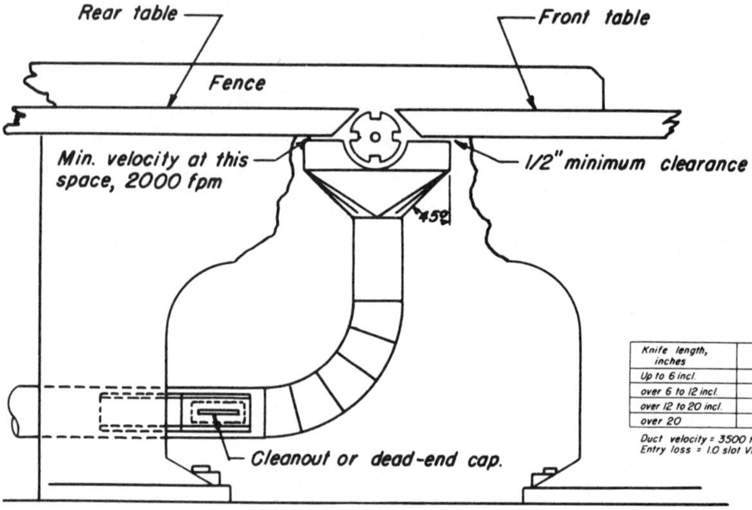

Knife length, inches	Exhaust volume, cfm
Up to 6 incl.	350
over 6 to 12 incl.	440
over 12 to 20 incl.	550
over 20	800

Duct velocity = 3500 fpm
Entry loss = 1.0 slot VP + 0.25 duct VP

Source: Committee on Industrial Ventilation, ACGIH. Adapted from VS-701 of *Industrial Ventilation.*

Figure 21 Horizontal Belt Sander Dust Collecting System

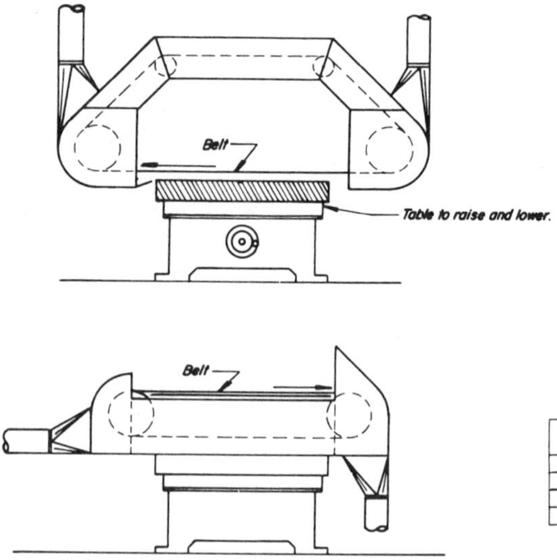

HORIZONTAL BELT SANDERS

Belt width, inches	Exhaust volume, cfm		
	Head end	Tail end	Total
Up to 6 incl.	440	350	790
over 6 to 9 incl.	550	350	900
over 9 to 14 incl	800	440	1240
over 14	1100	550	1650

Duct velocity = 3500 fpm
Entry losses = 0.40 VP for tapered take-off

Source: Committee on Industrial Ventilation, ACGIH. Adapted from VS-702 of *Industrial Ventilation.*

Figure 22 Disc Sander Dust Collecting System

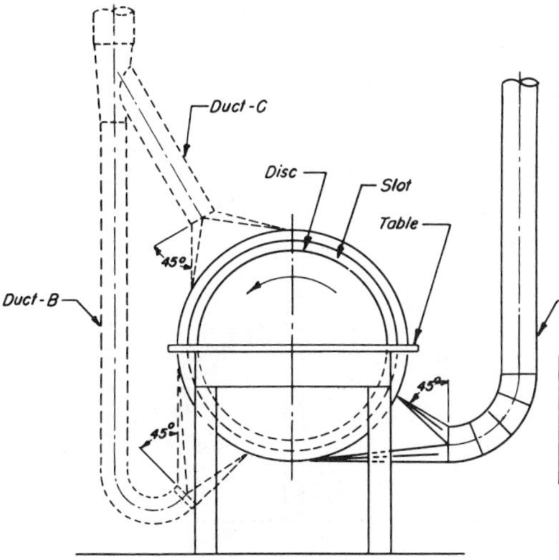

Disc diameter, inches	Total exhaust volume cfm	Applies to duct
Up to 12 incl.	350	A
over 12 to 18 incl.	440	A
over 18 to 26 incl.	550	A
over 26 to 32 incl.	700*	A-B
over 32 to 38 incl.	900*	A-B
over 38 to 48 incl.	1250**	A-B-C

* Two bottom branches.
** One top and two bottom branches.

Duct velocity = 3500 fpm
Entry loss: Depends on hood design.
1.0 slot VP + 0.25 duct VP

Source: Committee on Industrial Ventilation, ACGIH. Adapted from VS-703 of *Industrial Ventilation.*

Figure 23 Band Saw Dust Collecting System

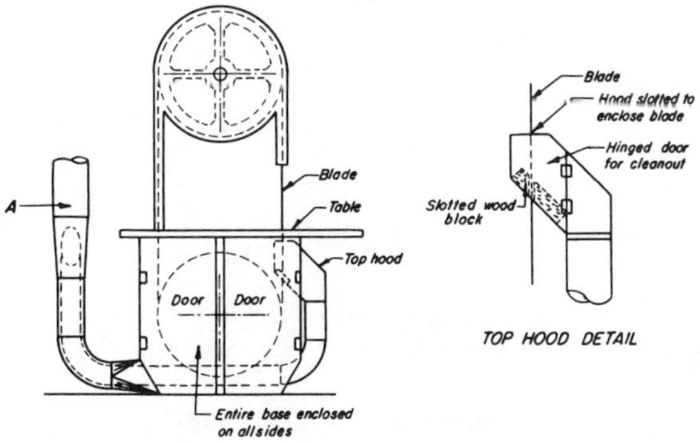

Blade width, inches	Exhaust volume, cfm		
	Bottom	Top	Total
Up to 2	350	350	700
over 2 to 3	350	550	900
over 3 to 4	550	800	1350
over 4 to 6	550	1100	1650
over 6 to 8	550	1400	1950

Duct velocity = 3500 fpm
Entry loss = 1.75 VP in duct riser (Point A)

Source: Committee on Industrial Ventilation, ACGIH. Adapted from VS-706 of *Industrial Ventilation.*

Figure 24 Table Saw Dust Collecting System

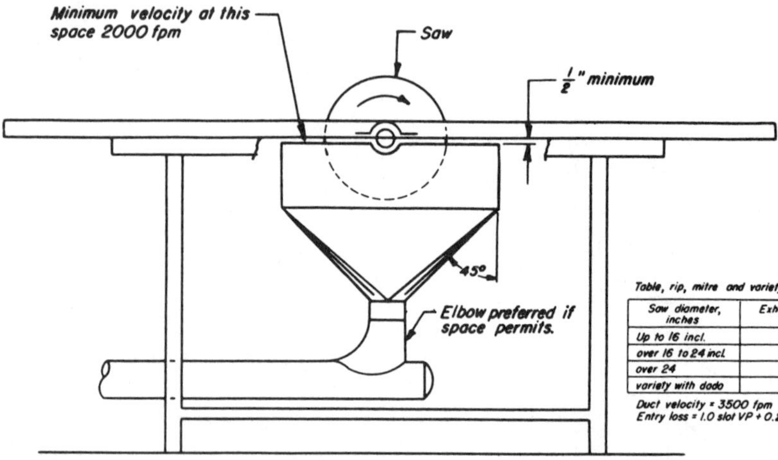

Table, rip, mitre and variety saws.

Saw diameter, inches	Exhaust volume, cfm
Up to 16 incl.	350
over 16 to 24 incl.	440
over 24	550
variety with dado	550

Duct velocity = 3500 fpm
Entry loss = 1.0 slot VP + 0.25 duct VP

Source: Committee on Industrial Ventilation, ACGIH. Adapted from VS-708 of *Industrial Ventilation.*

Figure 25 Radial Saw Dust Collecting System

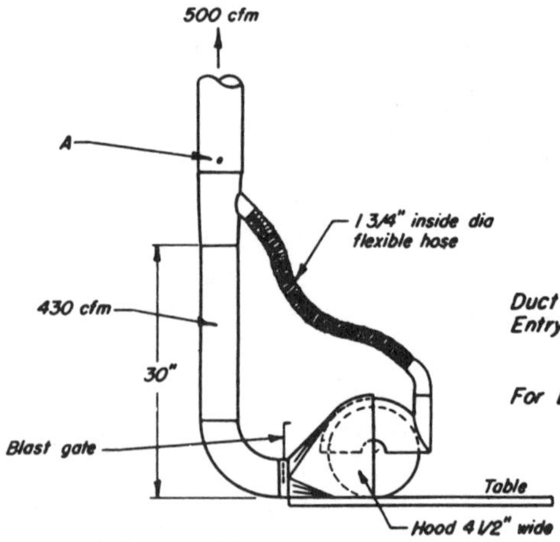

Duct velocity = 3500 fpm
Entry loss = 3.5 VP in duct riser (Point A)

For booth enclosure, see VS-401

Source: Committee on Industrial Ventilation, ACGIH. Adapted from VS-709 of *Industrial Ventilation.*

17 Movable exhaust system: welding

Movable exhaust systems are useful in situations where you need to be able to move an exhaust hood from one location to another, such as when you weld. Commercial movable exhaust systems for welding are readily available. Many consist of a flexible duct and hood attached to a portable cart which contains an electrostatic precipitator or fabric filter. Their major disadvantage is that they usually recirculate the exhaust air back into the workspace after the welding fumes have been removed by the filter or electrostatic precipitator. Recirculation systems like these cannot be recommended because they require frequent maintenance and because they do not remove other contaminants such as ozone and nitrogen dioxide which also can be produced during welding. If such units are equipped for outside exhaust, then they are acceptable.

In this instance it was decided to use a flexible exhaust system consisting of a plain opening hood, attached to six feet of flexible ducting, which in turn is attached to rigid metal ducting that runs horizontally for twenty-five feet to a window where the fan is located. The flexible ducting is supported by a pulley system. (See Figure 26.)

The hood consists of a plain opening with a flange attached for higher efficiency. Hoods with commercially available tapered cones also could be used. It was decided to place the hood six inches or less from the point of welding so as to avoid very high air capacities.

The simplified formula for calculating the capture air capacity of a plain opening hood is:

$$Q = V\,(10X^2)$$

where V = capture velocity in fpm

X = distance from source to hood in feet

Using a capture velocity V of 100 fpm, and a distance X of half a foot you will get:

$$Q = 100\ (10 \times (\tfrac{1}{2})^2)\ \text{cfm}$$
$$= 250\ \text{cfm}$$

With a flange, the efficiency will be greater.

Ducts

It is assumed that the air velocity in the flexible ducting is a maximum of 2,000 fpm because higher velocities will result in very high frictional losses. Using the formula:

$$Q = AV$$

where $Q = 250$ cfm
$V = 2{,}000$ fpm

you get a minimum duct diameter of 0.125 square feet. Consulting Table 12 of Appendix B, you will see that this falls between duct diameters of 4½ inches (.1105

Figure 26 **Movable Exhaust System**

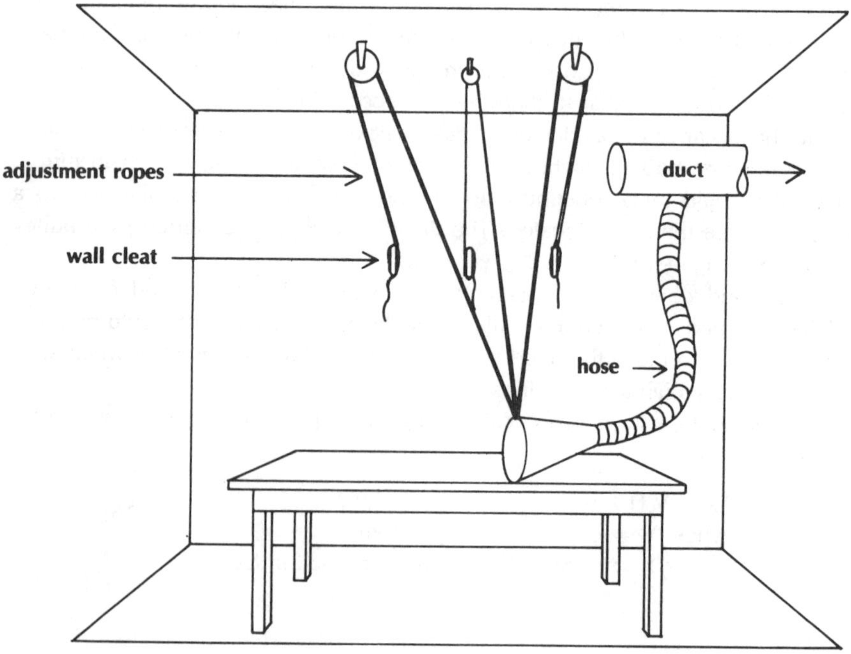

square feet) and five inches (.1364 square feet). Therefore you should choose the five-inch flexible duct and recalculate the duct velocity at 1,830 fpm.

Use the same duct size for the rigid ducting.

It is important that there be no kinks in the flexible hosing. Otherwise resistance would be increased enormously. In fact the air flow might decrease to nothing. In this example, assume that the flexible hosing has two $1\frac{1}{2}$-inch diameter forty-five-degree bends before it joins the rigid ducting. The rigid ducting has a straight run of twenty-five feet to the fan. Discharge is out the window through a short length of duct.

System resistance

The tables in Appendix B do not give values for calculating the resistance of flexible ducting. However, as a rule of thumb, you can assume that the resistance of flexible ducting is twice that of rigid ducting. Therefore, the calculations for the resistance of the flexible ducting can be made by proceeding as you would for rigid ducting, then using double the value of the resistance for 100 feet of galvanized ducting.

The system resistance then would be the sum of the hood resistance, flexible duct resistance, rigid duct resistance, and the discharge. If this is done as in the example in Appendix B, you will get:

Component		*Resistance (inches of water)*
Flanged hood		0.31
Flexible ducting		0.36
Rigid ducting		0.30
Exit loss (assumed)		0.10
	Total	1.07

Fan selection

The fan static pressure is calculated at 0.86 inches of water. (See Appendix B.) This is about 7/8 inches of water.

Knowing the fan static pressure and air capacity, you can consult manufacturers' catalogs to choose a fan. In this instance either a forward curve or backwardly inclined centrifugal fan would be appropriate.

A direct drive, forward curved fan, single-phase, 115 volts, with an air capacity of 380 cfm at 7/8-inch static pressure is available, rather than the calculated 250 cfm. To be sure of its effectiveness, a few more calculations are in order. The higher cfm would be even more effective in capturing the welding fumes, but the question is how the higher cfm affects the resistance. To maintain about 2,000 fpm duct velocity, six-inch ducting is needed instead of five-inch. Recalculating the resistance based on 380 cfm and six-inch ducting, you will find that the system resistance is now 1.05 inches instead of 1.07 inches, and the fan static pressure is 0.82 inches instead of 0.86 inches, using the new inlet diameter of six inches. You can ignore the small difference between the duct diameter of six inches and the fan inlet of 4 3/4 inches, since the difference is negligible. Therefore you can use this fan to deliver 380 cfm at 7/8 inches of fan static pressure. Since this is a direct drive fan, the motor comes as part of the unit.

The final fan data then are:

Type: forward-curved centrifugal fan
Size: BF9
Air capacity: 380 cfm
Fan static pressure: 7/8 inches of water
Fan speed: 3,400 RPM
Motor: 1/3 HP

Appendixes

APPENDIX A

Codes and regulations

The United States government, states, and municipalities issue regulations controlling toxic substances, flammable and combustible materials, and ventilation systems. They are intended to protect people's health, life, and property. Many are based on voluntary standards written by various organizations, such as the American National Standards Institute (ANSI), the National Fire Protection Association (NFPA), and the American Conference of Governmental Industrial Hygienists (ACGIH).

This appendix lists some of the federal Occupational Safety and Health Administration (OSHA) standards which may apply to the processes you use and to your ventilation system. Also included are some of the voluntary standards upon which the OSHA standards and many state and local codes are based. This list is not necessarily all-inclusive. Check with appropriate agencies in your area for specific requirements. Also check your lease (if you rent space) and any insurance policies (workers' compensation, fire, loss, damages, liability) for requirements concerning ventilation systems and the use of chemical substances.

Since some codes are revised every few years, you must determine if local codes refer to specific dates. In some instances a local code may require standards based upon a different code's "latest issue."

OSHA standards

The Occupational Safety and Health Administration, U.S. Department of Labor, issues health and safety standards for the workplace. The General Industry Standards 29 CFR 1910. are available from your local OSHA office.

If you are an employee or employer covered by the Occupational Safety and Health Act or if you work for yourself, you should be familiar with OSHA standards for toxic substances, flammable and combustible materials, and for ventilation systems. Adhering to the standards fulfills your obligation (if any) to the federal law and also assures you that your operation is safely controlled.

Specific standards for toxic substances are found in 1910.1000-1046. They include the Permissible Exposure Limits (PELs) for toxic substances and additional

Some specific OSHA standards are as follows.

1910.94 Ventilation: abrasive blasting; grinding, polishing, and buffing operations; spray finishing operations; and open surface tanks.

1910.106 Flammable and combustible liquids.

1910.107 Spray finishing using flammable and combustible materials.

1910.108 Dip tanks containing flammable or combustible liquids.

1910.1000–1046 Permissible Exposure Limits (PELs) for toxic substances and additional standards for highly toxic and carcinogenic chemicals.

Other federal agencies

Airborne emission standards for chemical pollutants are set by the United States Environmental Protection Agency (EPA). However, artists' studios, schools, and small shops are not likely to discharge contaminants in sufficient amounts to fall under EPA regulations. There may be local codes that regulate what, how much, and in what manner chemical pollutants may be discharged into the environment. Check with appropriate agencies in your area.

The Consumer Product Safety Commission (CPSC) is the federal agency that administers the Federal Hazardous Substances Act and the Consumer Product Safety Act. These acts regulate the banning and labeling of hazardous consumer products.

State and local codes

Various state and local government agencies also have standards for toxic substances, flammable and combustible materials, and ventilation systems. Check with appropriate agencies in your area such as the fire department, labor department, building department, air resources agency, and health department for legal requirements regarding safe operation of the processes you use.

Many local and state agencies have adopted OSHA standards or the voluntary standards set by such organizations as the American National Standards Institute (ANSI) and the National Fire Protection Association (NFPA). The following list includes some of the voluntary standards cited in OSHA references. These standards are often the basis for local codes as well. Whether you are bound by legal requirements or not, it is a good idea to be familiar with, and to follow these standards.

VENTILATION

ANSI Z9.1 (1951) Safety Code for Ventilation and Operation of Open Surface Tanks

ANSI Z9.2 (1971) Fundamentals Governing the Design and Operation of Local Exhaust Ventilation

American Conference of Governmental Industrial Hygienists (ACGIH), *Industrial Ventilation: A Manual of Recommended Practice,* updated regularly (the single best source about industrial ventilation practices)

FLAMMABLE AND COMBUSTIBLE MATERIALS

NFPA 30 Flammable and Combustible Liquids Code

NFPA 33 Standard for Spray Finishing Using Flammable and Combustible Material

NFPA 34 Standard for Dip Tanks Containing Flammable or Combustible Liquids

INSTALLATION OF ELECTRICAL EQUIPMENT

NFPA 70 National Electric Code (NEC)—use latest issue, unless local codes reference a specific code. The National Electric Code was developed and published by NFPA. The provisions of NEC Articles 500 through 503 apply to hazardous locations where explosion-proof and other special electrical equipment are required.

ADDRESSES

American Conference of Governmental
Industrial Hygienists (ACGIH)
6500 Glenway Avenue, Bldg. D-5
Cincinnati, OH 45211

American National Standards Institute (ANSI)
1430 Broadway
New York, NY 10036

American Society of Heating, Refrigerating
and Air Conditioning Engineers
(ASHRAE)
1791 Tullie Circle N.E.
Atlanta, GA 30329

Air Movement and Control Association (AMCA)
30 West University Drive
Arlington Heights, IL 60004

Consumer Product Safety Commission (CPSC)
Washington, DC 20207

National Fire Protection Association (NFPA)
Batterymarch Park
Quincy, MA 02269

Occupational Safety and Health Administration (OSHA)
U.S. Department of Labor
Washington, DC 20210

Environmental Protection Agency (EPA)
401 M Street S.W.
Washington, DC 20460

Additional information and assistance can be obtained from the Center for Occupational Hazards at:

Center for Occupational Hazards
5 Beekman Street
New York, NY 10038
212/227-6220

APPENDIX B

Calculating system resistance and selecting a fan

In order to select a proper fan for a local exhaust system, you need to know the system's resistance. Resistance is calculated by adding the separate resistances for (1) the hood, (2) ductwork, (3) air cleaner (if any), and (4) discharge from the fan.

The example of the canopy hood discussed in Chapter 13 is used here to illustrate these calculations. This hood has an air capacity of 1,250 cfm, and will be used with twelve-inch ducting with a duct velocity of 1,590 fpm. Duct diameter is determined from the formula $Q = AV$; Table 12 gives the duct diameter for various cross-sectional areas. As shown in Figure 27, there are two feet of ducting, then a ninety-degree elbow, another eight feet of ducting, another ninety-degree elbow, and finally five feet of straight ducting leading into the exhaust fan. The fan outlet is equipped with a back damper.

The following calculations are based on material in *Industrial Ventilation: A Manual of Recommended Practice*.

1. *Hood Resistance*. The hood resistance or hood static pressure comes from the hood entry loss caused by air turbulence and by the acceleration of the air entering the duct. The hood resistance SP_h is then $SP_h = h_e + VP_d$, where h_e is hood entry loss, and VP_d is the velocity pressure in the duct. Values of the hood entry loss (h_e) for various hoods are listed in Table 13.

For a round canopy hood with an included angle of 90°, $h_e = 0.15\ VP_d$.

The total hood resistance for the canopy hood is:

$SP_h = h_e + 1\ VP_d = 0.15\ VP_d + 1\ VP_d = 1.15\ VP_d$

Values of velocity pressure in inches of water can be found for different duct velocities from the figures in Table 14. These represent values calculated for standard air, which is applicable for conditions near room temperature and for

Figure 27 Canopy Hood Exhaust System

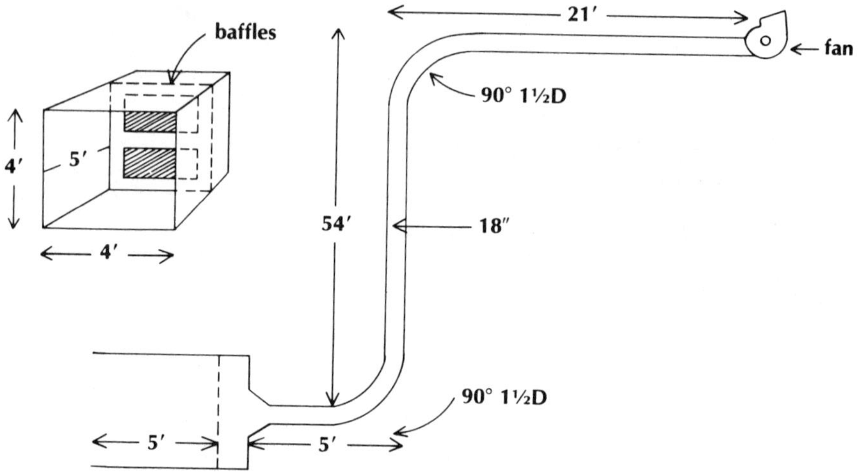

Table 12

Cross-Sectional Areas of Circular Ducts

Duct diameter (inches)	Area (square feet)	Duct diameter (inches)	Area (square feet)
1	0.0054	9	0.4418
1½	0.0123	9½	0.4923
2	0.0218	10	0.5454
2½	0.0341	11	0.6600
3	0.0491	12	0.7854
3½	0.0668	13	0.9218
4	0.0873	14	1.069
4½	0.1105	15	1.227
5	0.1364	16	1.396
5½	0.1650	17	1.576
6	0.1964	18	1.767
6½	0.2305	19	1.969
7	0.2673	20	2.182
7½	0.3068	21	2.405
8	0.3491	22	2.640
8½	0.3940	23	2.885
		24	3.142

air pressures up to about 2,000 feet of altitude. For the twelve-inch duct in the example, you will find the duct velocity to be 1,590 fpm. Therefore, the velocity pressure VP_d is 0.16 inches of water.

$SP_h = 1.15\ VP_d = 1.15 \times 0.16$ inches of water

$SP_h = 0.18$ inches of water

Note that hoods bought from manufacturers (that is, spray booths) may have their hood loss specified by their manufacturers.

2. *Duct Resistance.* The duct resistance or duct static pressure is calculated by adding the resistances of the straight runs of duct to the resistances of the other duct components such as elbows, joining of two ducts, changes in duct diameter, etc. Only elbows are considered here. More complicated duct systems can be found in *Industrial Ventilation: A Manual of Recommended Practice* and should be calculated by an engineer.

Duct resistance for galvanized standard duct is calculated by: (1) obtaining from Table 15 the resistance of elbows in terms of the number of equivalent feet of straight duct; (2) adding this equivalent duct length to the number of feet of straight ducting to obtain the total equivalent duct length; (3) obtaining from Table 15 the resistance of that diameter of ducting in inches of water per 100 feet of duct; and (4) multiplying the number of equivalent feet of ducting by the duct resistance per 100 feet to obtain the duct resistance.

Table 13

Hood Entry Losses for Various Hood Types

Hood types	h_e hood entry loss
Plain opening	$0.93\ VP_{duct}$
Plain opening with flange	$0.49\ VP_{duct}$
Canopy hood, rectangular (90° included angle*)	$0.25\ VP_{duct}$
Canopy hood, round (90° included angle*)	$0.15\ VP_{duct}$
Slot hood	$1.78\ VP_{duct}$
Spray booth (without baffles)	$0.50\ VP_{duct}$
Grinder hood	$0.65\ VP_{duct}$

* Canopy hood with included angles other than 90° will have different hood entry losses.

Table 14

Velocity Pressures for Different Velocities—Standard Air

$V = 4005 \sqrt{VP}$

V = Velocity FPM

VP = Velocity pressure, inches of water

VP	V	VP	V	VP	V	VP	V	VP	V	VP	V
0.01	400	0.52	2888	1.03	4064	1.54	4970	2.05	5734	3.10	7051
0.02	566	0.53	2916	1.04	4084	1.55	4986	2.06	5748	3.20	7164
0.03	694	0.54	2943	1.05	4103	1.56	5002	2.07	5762	3.30	7275
0.04	801	0.55	2970	1.06	4123	1.57	5018	2.08	5776	3.40	7385
0.05	896	0.56	2997	1.07	4142	1.58	5034	2.09	5790	3.50	7492
0.06	981	0.57	3024	1.08	4162	1.59	5050	2.10	5804	3.60	7599
0.07	1060	0.58	3050	1.09	4181	1.60	5066	2.11	5817	3.70	7704
0.08	1133	0.59	3076	1.10	4200	1.61	5082	2.12	5831	3.80	7807
0.09	1201	0.60	3102	1.11	4219	1.62	5098	2.13	5845	3.90	7909
0.10	1266	0.61	3127	1.12	4238	1.63	5114	2.14	5859	4.00	8010
0.11	1328	0.62	3153	1.13	4257	1.64	5129	2.15	5872	4.10	8109
0.12	1387	0.63	3179	1.14	4276	1.65	5144	2.16	5886	4.20	8208
0.13	1444	0.64	3204	1.15	4295	1.66	5160	2.17	5899	4.30	8305
0.14	1498	0.65	3229	1.16	4314	1.67	5175	2.18	5913	4.40	8401
0.15	1551	0.66	3254	1.17	4332	1.68	5191	2.19	5927	4.50	8496
0.16	1602	0.67	3279	1.18	4350	1.69	5206	2.20	5940	4.60	8590
0.17	1651	0.68	3303	1.19	4368	1.70	5222	2.21	5954	4.70	8683
0.18	1699	0.69	3327	1.20	4386	1.71	5237	2.22	5967	4.80	8774
0.19	1746	0.70	3351	1.21	4405	1.72	5253	2.23	5981	4.90	8865
0.20	1791	0.71	3375	1.22	4423	1.73	5268	2.24	5994	5.00	8955
0.21	1835	0.72	3398	1.23	4442	1.74	5283	2.25	6008	5.10	9044
0.22	1879	0.73	3422	1.24	4460	1.75	5298	2.26	6021	5.20	9133
0.23	1921	0.74	3445	1.25	4478	1.76	5313	2.27	6034	5.30	9220
0.24	1962	0.75	3468	1.26	4495	1.77	5328	2.28	6047	5.40	9307
0.25	2003	0.76	3491	1.27	4513	1.78	5343	2.29	6061	5.50	9392
0.26	2042	0.77	3514	1.28	4531	1.79	5359	2.30	6074	5.60	9477
0.27	2081	0.78	3537	1.29	4549	1.80	5374	2.31	6087	5.70	9562
0.28	2119	0.79	3560	1.30	4566	1.81	5388	2.32	6100	5.80	9645
0.29	2157	0.80	3582	1.31	4583	1.82	5403	2.33	6113	5.90	9728
0.30	2193	0.81	3604	1.32	4601	1.83	5418	2.34	6128	6.00	9810
0.31	2230	0.82	3625	1.33	4619	1.84	5433	2.35	6140	6.10	9891
0.32	2260	0.83	3657	1.34	4636	1.85	5447	2.36	6153	6.20	9972
0.33	2301	0.84	3669	1.35	4653	1.86	5462	2.37	6166	6.30	10052
0.34	2335	0.85	3690	1.36	4671	1.87	5477	2.38	6179	6.40	10132
0.35	2369	0.86	3709	1.37	4688	1.88	5491	2.39	6192	6.50	10210
0.36	2403	0.87	3729	1.38	4705	1.89	5506	2.40	6205	6.60	10289
0.37	2436	0.88	3758	1.39	4722	1.90	5521	2.41	6217	6.70	10366
0.38	2469	0.89	3779	1.40	4739	1.91	5535	2.42	6230	6.80	10444
0.39	2501	0.90	3800	1.41	4756	1.92	5550	2.43	6243	6.90	10520
0.40	2533	0.91	3821	1.42	4773	1.93	5564	2.44	6256	7.00	10596
0.41	2563	0.92	3842	1.43	4790	1.94	5579	2.45	6269	7.50	10968
0.42	2595	0.93	3863	1.44	4806	1.95	5593	2.46	6282	8.00	11328
0.43	2626	0.94	3884	1.45	4823	1.96	5608	2.47	6294	8.50	11676
0.44	2656	0.95	3904	1.46	4840	1.97	5623	2.48	6307	9.00	12015
0.45	2687	0.96	3924	1.47	4856	1.98	5637	2.49	6320	9.50	12344
0.46	2716	0.97	3945	1.48	4873	1.99	5651	2.50	6332	10.00	12665
0.47	2746	0.98	3965	1.49	4889	2.00	5664	2.60	6458	11.00	13283
0.48	2775	0.99	3985	1.50	4905	2.01	5678	2.70	6581	12.00	13874
0.49	2804	1.00	4005	1.51	4921	2.02	5692	2.80	6702	13.00	14440
0.50	2832	1.01	4025	1.52	4938	2.03	5706	2.90	6820	13.61	14775
0.51	2860	1.02	4045	1.53	4954	2.04	5720	3.00	6937	14.00	14986

Source: Committee on Industrial Ventilation, ACGIH.

Calculations for the example are as fol!ows:

Determine the type and number of duct components. For example, elbows are used in the ductwork whenever a change in direction is needed. Elbows usually are available in 30-, 45-, 60-, and 90-degree bends with a centerline radius (rate of turn) of 1.5 diameters (D) of the duct. The minimum recommended centerline radius is 1.5 D for exhausting vapors and gases; 2.0 D and 2.5 D elbows are often used for heavy dust loadings.

This system has two ninety-degree elbows with a diameter of twelve inches and a centerline radius of 1.5 D. For this elbow the equivalent duct length is twenty-five feet. Therefore the equivalent length of the two elbows is fifty feet.

The total length of straight ducting is fifteen feet. Therefore the total equivalent duct length is sixty-five feet. Table 16 plots four variables: friction loss in inches of water per 100 feet of duct, duct diameter, duct velocity, and air capacity (Q) in cfm in the duct. If you know two of the variables, you can read the other two from the chart. To find the resistance per 100 feet of duct you must know the duct diameter and either the cfm or the duct velocity.

Table 15

Equivalent Resistance in Feet of Straight Duct

Duct D	90° elbow* (1.5D centerline radius)	Duct branch entry (45° angle of entry)
3"	5	3
4"	6	5
5"	9	6
6"	12	7
7"	13	9
8"	15	11
10"	20	14
12"	25	17
14"	30	21
16"	36	25
18"	41	28
20"	46	32

* for 60° elbows–equivalent resistance is 0.67 × equivalent resistance for 90°
45° elbows–equivalent resistance is 0.50 × equivalent resistance for 90°
30° elbows–equivalent resistance is 0.33 × equivalent resistance for 90°

Source: Committee on Industrial Ventilation, ACGIH. Adapted from Fig. 6–11 of *Industrial Ventilation.*

In this example there is a twelve-inch duct and an air flow of 1,250 cfm. Look at Table 15. The intersection of the two lines representing these values occurs at a friction loss or resistance of 0.3 inches of water per 100 feet of duct. feet of duct.

Duct resistance = equivalent duct length × resistance/100 feet
= 65 feet × 0.3 inches of water/100 feet
= 0.20 inches of water.

3. *Air Cleaners.* The manufacturer should provide you the static pressure (resistance) of the air cleaner. You must note, however, that dust collectors' resistance shoots up when they become full or clogged. To keep resistance in line with design, most dust collectors should be changed or cleaned regularly. Some filter housings have built-in manometers to measure resistance through the filter. Usually when an increase in static pressure of 1/4 inch of water across the filter is noted, it is time to replace the filter. This example, however, includes no air cleaner.

4. *Discharge.* In this instance, at least two equivalent diameters of ductwork (at least three feet) are required at the discharge end of the fan to make sure the fan works with maximum effectiveness. Many other systems described in this book use a back draft damper (also called an outlet shutter or gravity damper). The total resistance of this damper and ductwork is assumed to be 0.1 inches of water.

5. *System Resistance.* This is found by adding the separate resistances for the system components.

Component	*Resistance (inches of water)*
Canopy hood	0.18
Ducting	0.20
Air cleaner (none)	0.0
Back damper	0.10
Total	0.48

Therefore the canopy exhaust system in this example has a resistance or static pressure of 0.48 inches of water.

After calculating system resistance, the next step is selecting a fan. To make your choice, you need to know: (1) air capacity, (2) fan static pressure, (3) type of contaminant, such as gases, vapors, fumes, dusts, etc., (4) fan speed, (5) fan brake horsepower, and (6) type of electrical service in your studio or shop. For the canopy exhaust system in this example, you know the following:

Air capacity, Q = 1,250 cfm
Duct diameter = 12 inches
Duct velocity pressure = 0.16 inches of water
System resistance = 0.48 inches of water
Form of contaminant: gases, steam, metal fumes, heated air

Table 16

Friction Loss in Inches of Water per 100 Ft

Air in Straight Ducts for Volumes of 10 to 2000 Cfm

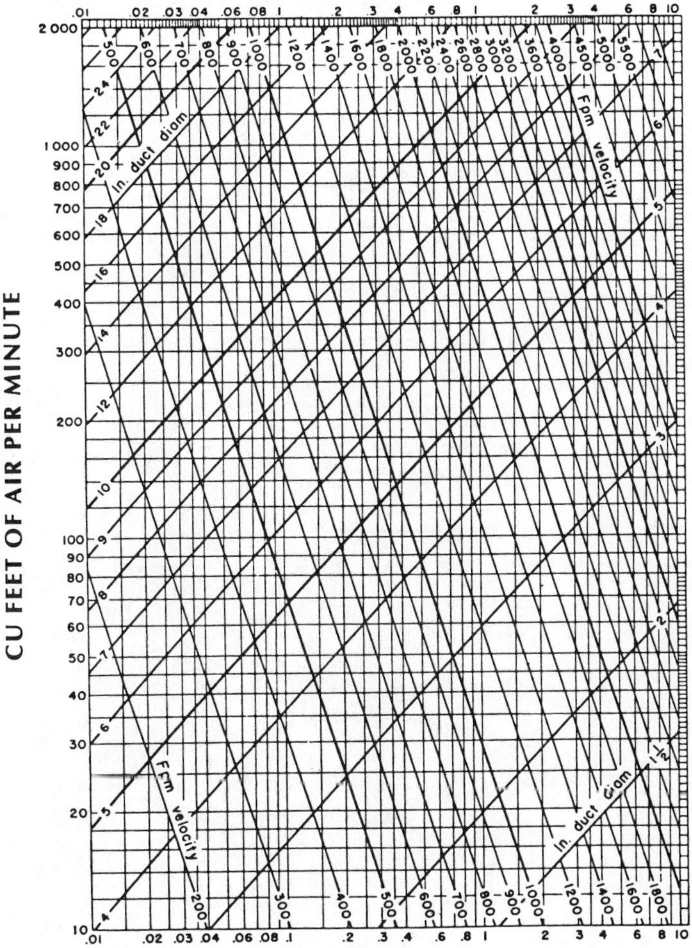

Based on Standard Air of 0.075 pound per cubic foot density flowing through average clean round galvanized metal ducts having approximately 40 joints per 100 feet. *Caution: Do not extrapolate below chart.* For proprietary duct, obtain data from manufacturer.

Source: 1972 Fundamentals Volume, ASHRAE Handbook, page 446. Reproduced by permission of American Society of Heating, Refrigerating and Air-Conditioning Engineers.

Now, review the fan features in Table 10. You will see that either a forward curved centrifugal fan or a backwardly inclined centrifugal fan can be used. A backwardly inclined centrifugal fan has been selected in this instance because it can handle light metal fumes and it has non-overloading characteristics. A fan diameter of 12 1/4 inches (called a 122 fan) can be used here, but other fan choices, too, are appropriate.

Next you calculate the fan static pressure, brake horsepower, and fan speed for a 122 backwardly inclined centrifugal fan, using a fan manufacturer's catalog.

1. *Fan static pressure*. The fan static pressure is calculated from the formula $SP_f = SP_{system} - VP_{fan\ inlet}$

If you choose a fan with the same inlet diameter as the duct (12 inches), then the velocity pressure at the fan inlet equals the duct velocity pressure (0.16 inches of water). The system resistance is 0.48 inches of water.

$$SP_f = 0.48 \text{ inches} - 0.16 \text{ inches}$$
$$= 0.32 \text{ inches of water}$$

Since fan static pressures usually are rated in multiples of 1/8 inches of water, choose a fan static pressure of 0.375 or 3/8 inches of water. You should increase rather than decrease the fan static pressure in order to have a safety factor.

Table 17

Fan Rating Table for Size 122 Backwardly Inclined Centrifugal Fan

CFM	OV	¼″ SP RPM	¼″ SP BHP	½″ SP RPM	½″ SP BHP
504	600	774	0.03	985	0.07
672	800	893	0.05	1067	0.09
840	1000	1019	0.08	1183	0.12
1008	1200	1156	0.12	1305	0.16
1176	1400	1306	0.16	1432	0.22
1344	1600	1460	0.23	1564	0.29
1512	1800	1618	0.30	1711	0.37

CFM is cubic feet of air per minute. BHP is the brake horsepower.
RPM is revolutions per minute. OV is a discharge or outlet velocity in feet/minute.

Source: AVCO Industries. Reproduced by permission.

2. *Brake horsepower*. Now you need to know what horsepower motor will deliver 1,250 cfm of air with this fan. The fan discussion in Chapter 10 recommends that a motor's horsepower should be at least twenty percent greater than brake horsepower. You usually cannot read the BHP directly from a fan rating table— and that is the case in this example. So you must interpolate between the given

values. Study Table 17, which is a fan rating table for a size 122 backwardly inclined centrifugal fan.

You will have to interpolate for 1,250 cfm at 3/8 inch static pressure in order to determine the brake horsepower (see Table 18). There are three steps involved: (1) interpolating between 1/4-inch and 1/2-inch static pressure to obtain the brake horsepower A at 3/8-inch static pressure and 1,176 cfm; (2) interpolating between 1/4-inch and 1/2-inch static pressure to get the brake horsepower B at 3/8-inch static pressure and 1,344 cfm; and (3) interpolating between these two values to obtain the brake horsepower C at 3/8-inch static pressure and 1,250 cfm.

Table 18

Interpolation from Fan Rating Table

	¼" SP		⅜" SP		½" SP	
CFM	RPM	BHP	RPM	BHP	RPM	BHP
1176	1306	0.16	—	A	1432	0.22
1250			—	C		
1344	1460	0.23	—	B	1564	0.29

For the first interpolation, 3/8-inch SP is half the difference between 1/4-inch and 1/2-inch SP. Therefore the brake horsepower at 3/8-inch SP and 1,176 cfm

$$= \text{(BHP at 1/4-inch SP)} + 1/2\ \text{(BHP at 1/2-inch SP} - \text{BHP at 1/4-inch SP)}$$
$$= 0.16 + 1/2\ (0.22 - 0.16)$$
$$= 0.19$$

Similarly, the brake horsepower at 3/8-inch SP and 1,344 is interpolated to 0.26 BHP, and the final interpolation at 3/8-inch SP and 1,250 cfm gives a brake horsepower of 0.22.

3. *Fan speed.* Similar interpolations for the fan speed yields a fan speed at 3/8-inch SP and 1,250 cfm of 1,432 rpm. Therefore the principal fan data for selection and purchase of a fan for the canopy system example are:

Fan: 12 1/4-inch backwardly inclined centrifugal fan
cfm: 1,250
Fan static pressure: 3/8-inch water
Fan speed: 1,432 rpm
BHP: 0.22

In choosing the motor for a fan needing 0.22 BHP, remember that you need a motor with a horsepower at least twenty percent higher than the BHP, or at least 0.27 horsepower. Therefore either a 1/3 or 1/2 horsepower single-phase motor could be selected. The larger motor would be the safer choice, since it allows for greater adjustments and for design inaccuracies. Elements involved in mounting a fan are discussed in Chapter 10.

APPENDIX C

Suppliers

Ventilation systems and components are available from manufacturers, distributors, and local ventilation and sheet metal shops. Check your area's Yellow Pages under the headings: ventilation contractors, ventilation equipment, sheet metal work, air pollution control, spraying equipment, fans-industrial and commercial, and exhaust fans. Ask friends and associates for the name of a reputable supplier. If you draw a blank, then write to a few of the suppliers, get some estimates, and select a supplier who understands your needs and can supply you with good equipment at a fair price.

The following list of suppliers will give you an idea of what is available. When you plan your system, contact any of the suppliers for product descriptions and prices. This list is not all-inclusive and does not represent an endorsement by the publisher or authors.

Component Parts for Ventilation System: ducts, elbows, dampers, etc.

The Kirk and Blum Manufacturing Co.
3120 Forrer Street
Cincinnati, Ohio 45209
(513) 351-1400

United Sheet Metal Div.
United McGill Corp.
200 E. Broadway
Westerville, OH 43081
(614) 882-7401

Dust Collectors

Aget Mfg. Co.
1408 E. Church Street
P.O. Box 248
Adrian, Mich. 49221
(517) 263-5781

DCE Vokes Inc.
Plainview Plaza, Suite 900
10101 Linn Station Road
Jeffersontown, KY 40223
(502) 426-8190

Torit Division
Donaldson Company, Inc.
P.O. Box 1299
St. Paul, Minnesota 55440
(612) 887-3921

Fans

Acme Engineering & Mfg. Corp.
1820 North York Street
Muskogee, OK 74401
(918) 682-7791

Aerovent Inc.
1 Aerovent Drive
Piqua, OH 45356
(513) 773-4611

AVCO Industries Inc.
37-13 Greenpoint Avenue
Long Island City, N.Y. 11101
(212) 784-2792

Barry Blower
99 NE 77 Way
Minneapolis, MN 55432
(612) 571-5200

Buffalo Forge Co.
490 Broadway
Buffalo, N.Y. 14240
(716) 847-5121

Ilg Industries
Division of Carrier Corporation
2850 North Pulaski Road
Chicago, Illinois 60641
(312) 725-8016

New York Blower Co.
3155 Shields Ave.
Chicago, IL 60616
(312) 655-4881

Filters

American Air Filter Co.
215 Central Avenue
Louisville, KY 40277
(502) 637-0011

Balston Filter Products
703 Massachusetts Avenue
P.O. Box C
Lexington, Massachusetts 02173
(617) 861-7240

Barnebey Cheney
Columbus, OH 43216
(614) 258-9501

Cambridge Filter Corporation
P.O. Box 1255
7645 Seventh North Road
Syracuse, N.Y. 10201
(315) 457-1000

Portable Hoods (Welding Fume Extractors)

Ecology Tech
P.O. Box 1598
3521 Airport Road
Jonesboro, Arizona 72401
(501) 932-6262 Toll free: (800) 643-0000

Nederman Inc.
1051 Clinton Street
Buffalo, N.Y. 14206
(716) 849-0391

Ruemelin Mfg. Co. Inc.
3860 N. Palmer Street
Milwaukee, Wisconsin 53212
(414) 962-6500

United Air Specialists, Inc.
4440 Creek Road
Cincinnati, OH 45242
(513) 891-0400

Exair Division (Soldering)
I C Industries Inc.
P.O. Box 689
Altamonte Springs, Florida 32701
(305) 339-8298

Special Vacuum Cleaners

Hako Minuteman
111 South Route 53
Addison, Illinois 60101
(312) 627-6900

Lab Safety Supply
P.O. Box 1368
Janesville, Wisconsin 53547
(800) 356-0783

Nilfisk of America, Inc.
3244 Prospect Street, N.W.
Washington, D.C. 20007
(202) 338-9360

Spray Booths

DeVilbiss Company
Toledo, OH 43692
(419) 470-2169

Binks Manufacturing Company
9201 West Belmont Avenue
Franklin Park, Illinois 60131
(312) 671-3000

APPENDIX D

Bibliography

The list of publications below is intended to give the reader specific sources for information presented in the text and general references.

Air Movement and Control Association. "Fans and Systems," Part 1 of *AMCA Fan Application Manual*. Publication 201. Arlington Height, IL.; AMCA, 1979.

A. M. Best Company. *Best's Safety Directory*. 2 vols. Oldwick, NJ: A. M. Best Co., 1982.

American National Standards Institute. *American National Standard: Fundamentals Governing the Design and Operation of Local Exhaust Systems*. Z9.2-1979. New York: ANSI, 1980.

Committee on Industrial Ventilation, ACGIH. *Industrial Ventilation: A Manual of Recommended Practice*. 17th ed. Lansing, MI.: ACGIH, 1982.

Consumer Product Safety Commission. *Federal Hazardous Substances Act Regulations*. Federal Register: September 27, 1973.

McCann, Michael. *Artist Beware*. New York: Watson-Guptill, 1979.

National Fire Protection Association. *Blower and Exhaust System 1973*. NFPA 91. Quincy, MA.: NFPA, 1973.

National Fire Protection Association. *National Electrical Code*. Quincy, MA.: NFPA, 1984.

National Institute of Occupational Safety and Health. *The Industrial Environment—Its Evaluation and Control*. Washington, D.C.: Government Printing Office, 1973.

National Safety Council. *Accident Prevention Manual for Industrial Operations*. 8th ed. Chicago: NSC, 1981.

Sax, N. Irving. *Dangerous Properties of Industrial Materials*. 5th ed. New York: Van Nostrand-Rheinhold, 1979.

Stellman, Jeanne and Susan Daum. *Work Is Dangerous to Your Health*. New York: Vintage, 1973.

U.S. Department of Labor. *General Industry Occupational Safety and Health Standards*. 29CFR 1910. Federal Register: June 1981 (rev.).

APPENDIX E

Glossary

Air capacity (Q), also called air quantity and exhaust volume, is the volume of moving air per unit of time.

Air cleaners collect the contaminants (dusts, fumes, vapors, gases) from the air as the air passes through them. Air cleaners include air filters, cyclones, charcoal filters, electrostatic precipitators, air washers (scrubbers), etc.

Air pressure is the exertion of force by air.

Atmospheric pressure is the pressure exerted by the weight of the atmosphere on the earth. *Standard atmospheric pressure,* measured at sea level, is equal to the pressure exerted by 29.92 inches of mercury or 407.5 inches of water.

Static pressure is exerted by all air. It is the potential pressure exerted in all directions that must be overcome for air to move. Static pressure is either positive or negative in relation to atmospheric pressure. *Positive static pressure* causes air molecules to be packed together tightly so that they tend to burst open their container. *Negative static pressure* causes air molecules to thin out so that they tend to draw together and collapse their container. Static pressure is used to refer to resistance in the components of a ventilation system.

Velocity pressure is force exerted by air that is moving. It is exerted in the direction of flow and is always positive. Velocity pressure is related to velocity so that for any velocity a value for the velocity pressure can be determined. (See Appendix B's Table 14 for velocity pressure values and corresponding values of velocity.) Velocity pressure in the ducts of a ventilation system is used to compute pressure loss of air entering a hood.

Total pressure is the sum of the static pressure and the velocity pressure.

Air resistance is the hindrance to the flow of air caused by the friction of the air molecules against other air molecules or against ventilation components and by the turbulence that occurs when air flow changes speed or direction.

Air velocity is the speed of air in feet per minute (fpm). Air velocity varies throughout a ventilation system depending primarily upon the shape of the part through which it passes. Average velocity (V) must be known in order to determine the air capacity (Q) flowing through any one part of the system.

Capture velocity is the air velocity required to draw the contaminated air into the hood and away from the source of the contamination.

Duct velocity is the average air velocity in the duct.

Transport velocity, or conveying velocity, is the velocity in the ducts required to keep solid particles moving in the airstream. See Table 9 chart for minimum velocities.

Face velocity is the average air velocity across the opening of a local exhaust hood.

Plenum velocity is the average air velocity in a plenum. (See *plenum*.)

Slot velocity is the average air velocity through the slotted opening(s) of a slot hood. The slot velocity allows for uniform air distribution across the length of the slot opening.

Balancing. A ducted ventilation system which has multiple exhaust openings or air outlets must be designed so that the minimum required air capacity (Q) is provided to each hood or outlet. The air resistance of each component must be controlled either by permanent design of the system or by adjustable dampers installed in the ductwork. Whenever a change in the hoods or the ductwork is made, the whole system must be rebalanced.

Brake horsepower (BHP) is the horsepower required to operate a fan alone. It does not include frictional losses related to the motor and motor/fan connections. (It is not the motor horsepower.)

Centerline radius is the rate of turn in duct elbows. The minimum recommended centerline radius is 1.5 duct diameters (D); for heavy dusts, a more gradual turn of 2.0 D or 2.5 D may be necessary.

Damper is a movable plate installed in the ductwork to regulate air flow. A blast gate is a type of damper used in ventilation systems to balance air flow through various branches of the system.

Duct or ductwork is the part of the local exhaust ventilation system that carries the contaminated air from the hood to the discharge. It is measured in *actual length* for straight runs of duct and in *equivalent length* for elbows and other fittings.

Fan is an air-moving device. It creates the pressure differential needed for air to move.

Flange is a flat rim around the edge of a hood which allows only the air in front of the hood to be drawn in. It can be up to one diameter in width, but it need not be more than six inches wide.

Hood is the part of a local exhaust ventilation system which captures the contaminated air. The *hood face* is the plane of the opening through which the air is drawn.

Make-up air is air brought into the workspace to replace the air exhausted by the ventilation system.

Overload characteristic is the tendency of a fan motor to burn out from electrical overload. This tendency is most typical of the forward curved centrifugal fan.

PELs are the Permissible Exposure Limits set by the Occupational Safety and Health Administration, a federal agency. They are also eight-hour time-weighted average concentrations in air.

Plenum. A plenum is an air pressure equalization chamber. In a slot hood, the plenum serves as an air space behind the slot where the air velocity through the slot (or slots) is slowed before entering the ductwork.

Positive pressure. See *air pressure*.

Pressure. See *air pressure*.

Resistance. See *air resistance*.

RPM, revolutions per minute, is the speed of a fan measured by the number of rotations the fan wheel makes around the fan shaft each minute.

TLV (Threshold Limit Values) are levels of airborne contaminants that most people can be exposed to without adverse effects. They are established by ACGIH, which periodically revises and updates them, and are published annually.

TLV-TWA is the eight hour Time-Weighted Average TLV. *TLV-STELs* are Short-Term Exposure Limits set by the ACGIH. They are fifteen-minute average exposures. *Ceiling TLVs* are concentrations that are never to be exceeded, even momentarily.

Velometer is a trade name for a swinging vane anemometer produced by Alnor; used to measure air velocity.

Index